Animals & Men

Are we the only Humans?

Yeti - ape not bear; the Ohio dogman; Zoological oddities in *Forest and Stream* Magazine; WW North; news, reviews and more...

Contents

Typeset by Jonathan Downes,
Cover and Layout by SPiderKaT for CFZ Communications
Using Microsoft Word 2000, Microsoft Publisher 2000, Adobe Photoshop CS.
First published in Great Britain by CFZ Press

CFZ Press, Myrtle Cottage, Woolsery, Bideford, North Devon, EX39 5QR

© CFZ MMXV

978-1-909488-46-5

Faculty of the Centre for Fortean Zoology

EDITORIAL

Dear Friends,

Welcome to another issue of the world's longest standing cryptozoological magazine. Before we continue any further I must apologise for the lateness of this issue. By our standards it is not particularly late, but it should have been released before the end of June, and by the time that you read this it will be well into July.

The reason? A happy one for once.

Corinna and I (together with Mama-in-law) went on holiday at the end of last month, to a

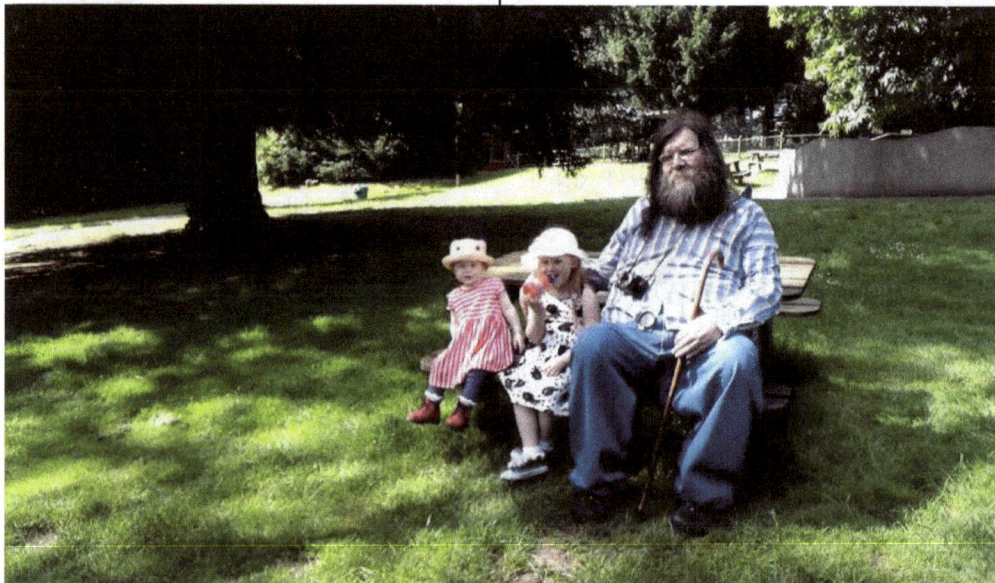

The Great Days of Zoology are not done!

family wedding, and then to Norwich to visit my younger stepdaughter Olivia and my two delightful grandchildren.

I did what Grandads are supposed to do, and bought the two little girls colourful hats and took them to the zoo at Thrigby Hall, where they got sticky from ice creams, had a delightful frisson of fear from the crocodiles, and giggled at the bright red buttocks of the Sulawesi crested macaques just as I did when I was their age and the monkeys were still called Celebes apes. So, we are running a few weeks late, and I hope that you will forgive me.

For me the highlight of the zoo visit (except of course for spending it with my beloved granddaughters) was the swamp house where I saw excellent exhibits of three species that I have never seen in Britain (two of which I don't think that I have ever seen at all).

These were Indopacific crocodiles (*C. porosus*), mugger crocodiles (*C palustris*) - one of which is pictured above - and *P pangasius* one of the species of Southeast Asian giant catfish, which though not quite as large as the critically endangered Mekong Giant Catfish can still reach a respectable three metres in length.

My son-in-law Aaron drooled over the splendid reticulated and Burmese pythons, but admitted that having them would have to wait until the little girls are considerably larger before it would be safe to even consider having a snake of that size.

19-21 August 2016
The Small
School, Hartland,
Devon

www.weirdweekend.org

I know that I saw Indopacific crocodiles at a zoo in Australia when I was a boy, but I cannot actually ever remember seeing muggers before, and I was not prepared for quite how impressive they are in the flesh. One of this species is, of course, the central character in Kipling's short story *The Undertakers* which can be found in the *Second Jungle Book* (1895) and is illustrated here by John Lockwood Kipling.

Thrigby Hall is exactly what a small to medium sized zoo *should* be like and puts some of the other establishments that we have visited over the years to shame.

A zoo should be a temple where one goes to worship at the altar of Mother Nature, not a cheap and tacky tourist attraction where visitors are divested of their money by a seemingly endless collection of rides and vending machines, and where the animals are plonked into uninspiring exhibits with no attempt to teach visitors anything about them.

And if you think that you know where I am talking about, you are probably right.

In our increasingly urbanised and sedentary society zoos are increasingly important in demonstrating the glory of the natural history of the world to successive generations of our species who are in danger of becoming completely divorced from it.

Thrigby Hall ticks all the right boxes and more. Seldom have I been more impressed.

Well done chaps,

Jon Downes (Director, CFZ)

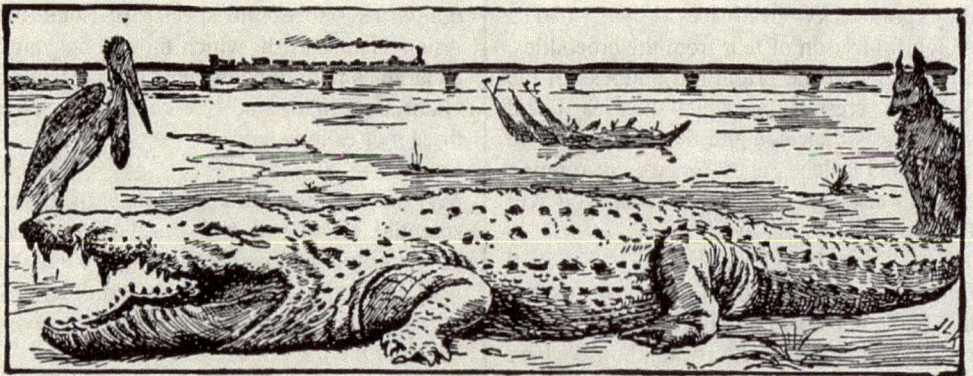

THE CENTRE FOR FORTEAN ZOOLOGY
www.cfz.org.uk

A LEGAL MATTER

Newsfile

New & Rediscovered

Ratting Tim Out

Manus Island's newest "detainee" may have been on the island hundreds of thousands of years. *Rattus detentus,* an ancient, isolated and previously unknown species of the genus Rattus – a rat – has been so named for the Latin "detained", "in reference to the isolation of ... Manus Island and to the recent use of the island to detain people seeking political and/or economic asylum in Australia".

The animal has been described for the first time, in the *Journal of Mammalogy,* by an international team of scientists including a former Australian of the Year, the mammalogist and palaeontologist Prof Tim Flannery. Before confirmation that detentus existed, Flannery said scientists had suspected there was a large rat endemic to the island. He said it had been exciting and "an immense privilege" to be able to discover and name the new species. "I've been looking for this rat for 30 years," he said.

SOURCE: http://tinyurl.com/hgysbrf

Silver Snake

In July of 2015 a team of scientists discovered a new species of boa during an expedition to a remote corner of the Bahamian Archipelago. They have named the new species the Silver Boa, *Chilabothrus argentum*. Significantly, this is the first new species of boa discovered in situ in the Caribbean since the 1940s. This new boa species is considered Critically Endangered, and is one of the most endangered boa species globally.

SOURCE: http://tinyurl.com/husvdue

First North American monkey

For decades one of the things which is cited every time that the subject of Bigfoot is raised is that there never have been any known primate species apart from us in North America. Now all that has changed.

Seven fossil teeth exposed by the Panama Canal expansion project are the first evidence of a monkey on the North American continent before the Isthmus of Panama connected it to South America 3.5 million years ago. A team including Carlos Jaramillo, staff scientist at the Smithsonian Tropical Research Institute (STRI), published this discovery online in the

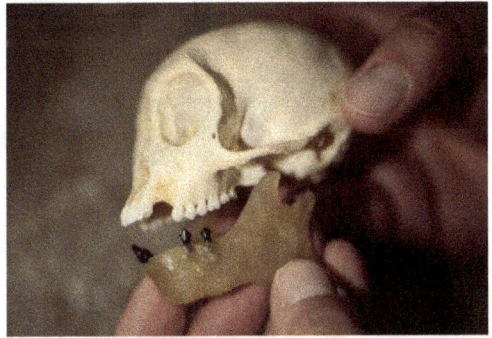

Placed in a wax jaw, fossil teeth belonging to *Panamacebus transitus* are compared with those of a modern female tufted capuchin, *Cebus apella*, in this picture courtesy of the Florida Museum of Natural History. Florida Museum of Natural History/Kristen Grace/

journal *Nature*. They named the new monkey species *Panamacebus transitus* in honour of Panama and the monkey's movement across the ancient seaway that divided North and South America.

The 21 million-year-old teeth were found in the Las Cascadas Formation during a five-year intensive fossil salvage project by field crews from STRI, the University of Florida and the New Mexico Museum of Natural History and Science. Most of the mammal groups represented in the Las Cascadas formation have North American origins, despite the fact that South America is much closer, supporting the idea that Central America and western Panama represented a long peninsula extending south from North America.

We would like to stress that whereas this is undoubtedly a major discovery, we are not going to extrapolate anything from it regarding putative North American man beasts.

SOURCE: **http://tinyurl.com/h5qxo92**

More Mouse Lemurs

Thanks to state-of-the-art genetic methods and expeditions to remote areas, researchers have just described three new mouse lemur species living in the South and East of Madagascar in a report published by the journal *Molecular Ecology*. Mouse lemurs are tiny, nocturnal primates, which are only seen in Madagascar, and all of them appear to be very similar, with their brown coat and large eyes. Various species can be individually identified by genetic techniques. However, determining the real differences between two populations is still a source of contention.

"By using new, objective methods to assess genetic differences between individuals, we were able to find independent evidence that these three mouse lemurs represent new species," study author Peter Kappeler, an expert from the German Primate Center, said in a news release. The study team, using the same genetic methods, was also able to affirm the status of 21 previously described species.

"The genetic techniques we used could facilitate species identification, thus also contributing to further new descriptions in other animal groups," Kappeler said. Just three years ago, the same researchers had identified two new mouse lemur species. Less than 20 years ago, only two types of mouse lemurs were known.

SOURCE: http://tinyurl.com/zhuos5p

Hawaiian fish surprise

Researchers in Hawaii have discovered three probable new species of fish while on an expedition in the protected waters of the Papahanaumokuakea Marine National Monument. In a statement released Wednesday, National Oceanic and Atmospheric Administration officials said divers collected two previously unknown species of fish and filmed a third.

NOAA's Randall Kosaki, the expedition's chief scientist, said the team collected the first specimens of male Hawaiian pigfish about 300 feet below the surface.

The scientists also observed significant coral mortality in the region that was the result of a mass bleaching event in 2014. Hawaii Institute of Marine Biology researcher John Burns said a 2015 trip found about 90 percent of the coral around Lisianski Island had died. This year, the team found that dead coral was covered in a green algae bloom.

Source: http://tinyurl.com/zf9gawf

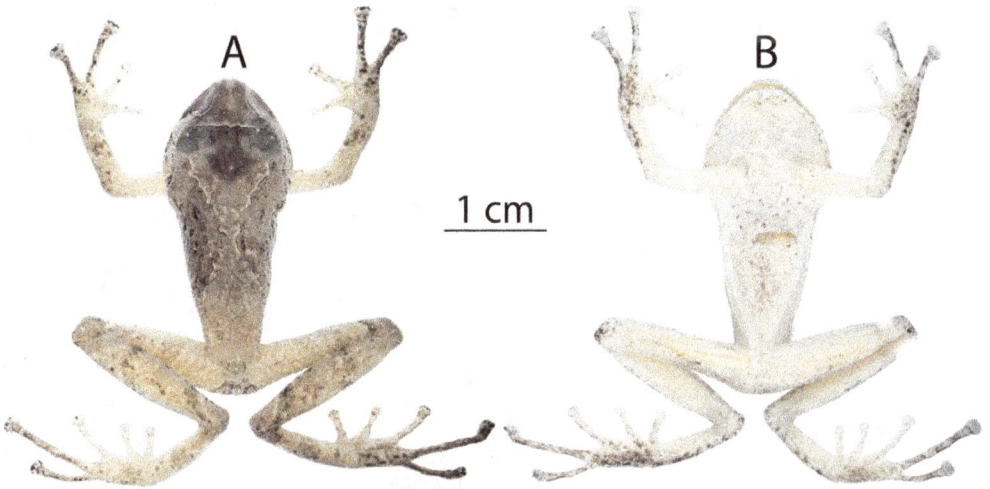

Preserved holotype of *Pristimantis yanezi* sp. n., QCAZ 46259, adult male, SVL = 29.8 mm. Dorsal (**A**), ventral (**B**) views.

Three rain frogs

Inaccessibility and mysticism surrounding the mist-veiled mountains of the central Andes make this region promising to hide treasures. With an area of 2197 km2, most of the Llanganates National Park, Ecuador, is nearly unreachable and is traversed only by

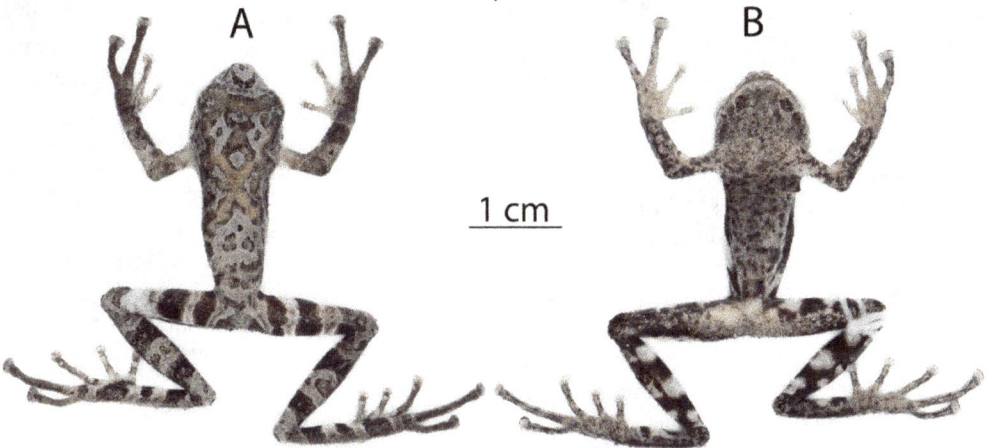

Preserved holotype of *Pristimantis llanganati* sp. n., QCAZ 46140, adult male, SVL = 27.1 mm. Dorsal (**A**), ventral (**B**).

Another Compendium of Batrachia

foot. However, fieldwork conducted by researchers from the Museo de Zoología at Catholic University of Ecuador resulted in the discovery of a more real and tangible gem: biodiversity.

Among other surprises, during their expeditions the researchers discovered two new species of rain frogs, formally named *P. llanganati* and *P. yanezi*. The new species are characterized by the spiny appearance typical of several species inhabiting montane forests. The study was published in the open access journal *ZooKeys*.

SOURCE: http://tinyurl.com/zp8gbgl

A third new species of rain frog has also been discovered. This new rain frog species has been described from Amazonian Peru and the Amazonian foothills of the Andes.

The frog, given the name *Pristimantis pluvialis,* was found by researchers from Southern Illinois University Carbondale, the University of Michigan, and the National University of San Antonio Abad of Cusco in Peru. The discovery is published in the open access journal *ZooKeys*.

SOURCE: http://tinyurl.com/jsb5dcr

Indian bush frog

Two new bush frog species have been discovered at the Silent Valley National Park, renowned as a repository of biodiversity. The species were found in a survey conducted by researchers and naturalists led by Anil Zachariah, veterinarian, and Robin Abraham, amphibian researcher, Kansas University, US. It is published in the latest issue of *Salamandra*, a German herpetological journal.

"The new species belongs to the bush frog genus Raorchestes and is located in the cloud forests of the Sispara and Thudukki sections of Silent Valley National Park," Dr. Zachariah told *The Hindu* . One species has been named *Raorchestes silentvalley* (above) in commemoration of the people's struggle to preserve this tract of tropical forest and grasslands in the south west of the Nilgiris, he said. The second species is named *Raorchestes lechiya* after the late Mr. Lechiyappan, a forest watcher who guided scientists in the park for rapid surveys, which helped in the region's eventual declaration as a national park, Mr. Abraham said.

SOURCE: http://tinyurl.com/h86x6bj

Another Compendium of Batrachia

Watch out Tokyo

A new goby has been discovered in the southern Caribbean. Living at depths greater than conventional SCUBA divers can access, yet too shallow to interest deep-diving submersibles, the fish will now be known under the common name of the Godzilla goby, referring to its reptilian appearance and proportions. Its discoverers Drs Luke Tornabene, Ross Robertson and Carole C. Baldwin, all affiliated with the Smithsonian Institution, have described the species in the open access journal *ZooKeys*. Formally called *Varicus lacerta*, the species name translates to 'lizard' in Latin and refers to the reptilian appearance of the fish. Its prime colours are bright yellow and orange, while the eyes are green.

The new goby also has a disproportionately large head and multiple rows of recurved canine teeth in each jaw. This is also why the research team has chosen the common name of the Godzilla goby. Apart from its lovely coloration, the new fish stands out with its branched, feather-like pelvic-fin rays and the absence of scales.

SOURCE: http://tinyurl.com/gmgsnh9

Centennial wasp

A species of wasp that is a natural enemy of a wood-boring beetle that kills black locust trees has been rediscovered, more than 100 years after the last wasp of this species was found. The discovery is significant because the wood-boring beetle, known as the locust borer, is considered a serious pest that has discouraged planting of black locusts, which played an important role in American history.

The trees, whose wood is strong, hard and extremely durable, helped build the Jamestown settlement and were featured prominently at George Washington's Mount Vernon.

The only previous known specimens of the wasp (*Oobius depressus*) date back to 1914 and were found in Morristown, Illinois. The problem with those specimens is that they were missing their heads and antennae, making them difficult to identify even by specialists of that wasp family, Encyrtidae. That led Serguei V. Triapitsyn, director of the UC Riverside Entomology Research Museum, and Toby R. Petrice, an entolomogist with the U.S.D.A. Forest Service Northern Research Station in Lansing, Mich., to search for new specimens.

This was not an easy task because eggs of locust borer, particularly ones parasitized by this wasp, are extremely difficult to find. Adults of the locust borer itself, on the other hand, are common in the Midwest in early fall because they feed on the pollen of goldenrod.

SOURCE: http://tinyurl.com/hth7oun

Thylacine

Sadly Not

Our dear friends/colleagues at CFZ Australia write:

Tis the season for Facebook pranks.

- Make sure you pretend you just stumbled on to the animal. Tick
- Make sure the animal doesn't react like a normal animal. Tick
- Pretend you do not know what the

Geoff Treloar added 2 new photos.
17 hrs · Scenes

Just seen this out hunting scared the shit out of me it ran past grunting did not get a shot just the camera

68

878 shares   27 comments

Deb McEwan
17 hrs

Justin Price Bugger Me what was it
17 hrs

Heather Treloar What is that?

animal is.. your just "putting it out there" .Tick

- Pretend you don't have any idea what the true value of the photos would be, if they were real. just give them away on Facebook. Tick

I don't know who Geoff Treloar is, but the

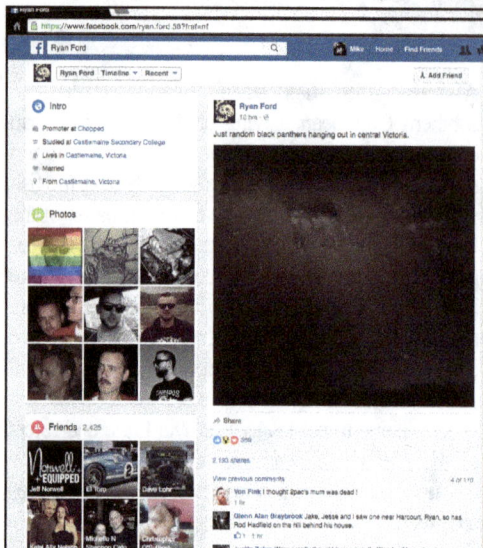

pictures were removed soon after. Interestingly it is the second crypto hoax on Facebook in recent weeks because there was a similar one purporting to be of a mystery big cat in Victoria which was solved by our own Tania Woodliffe.

Well done to CFZ Australia for documenting and pointing out these hoaxes. This is an important part of what we do even though it does lead the great unwashed in some quarters to claim we are naysayers and sceptics.

Pah!

SOURCES: http://tinyurl.com/zdah96a
http://tinyurl.com/zqsy8ek

Chupacabras/ Blue Dogs

Whiteface Wonder

A woman in Whiteface, Texas claimed to have seen a mythical chupacabras on a road in Hockley County. A Texas Parks and Wildlife Game Warden investigated the purported chupacabras sighting on Road 597 in Hockley County Saturday late afternoon (May 14). According to Texas Parks and Wildlife Game Warden Aaron Sims, a woman inside a convenience store in Whiteface (west of Levelland) said she had seen saw a creature in the road that "looked like the chupacabras you see in videos online."

Upon investigation, Sims confirmed to EverythingLubbock that the alleged chupacabras was merely a coyote, albeit a large one, suffering from mange. Photographs show the body distended by the gas of putrefaction, but not the strange pouches on the haunches which we have been investigating in some blue dog reports.

SOURCES:
http://kfyo.com/hockley-county-chupacabra/
http://tinyurl.com/j5oqjjj

This would never happen in OLD Hampshire

Some in New Hampshire are calling it the "zombie dog." Others have compared the wrinkled, emaciated creature seen trotting through a cemetery in Merrimack this week to the elusive chupacabras, a mythical beast said to roam parts of Latin America, feasting on goats. Merrimack Police Lieutenant Matthew Tarleton, who got a firsthand look at the patchy-haired, square-jawed animal with the long tail on Tuesday, said he had never seen anything like it before.

"To be honest, it looks like something out of a horror movie," he said in a telephone interview, as he prepared to set out to the area where the unidentified animal was last seen, to try to locate it. "I have seen some sick foxes and stuff, but this one — I don't know what it is — it's definitely unique-looking." Many thanks to Elizabeth Clem for sending this. It, once again, has the pads on its haunches which seem to be diagnostic of these particular creatures. Nobody is quite sure what they are, and they do appear to be largely diseased, but READ MY LIPS they are NOT chupacabras.

SOURCE: http://tinyurl.com/gwk2gtx

In late June pitchfork wielding villagers in Ukraine claimed to have killed a chupacabras. The creature – which is said to suck the blood of its victims – was reportedly killed in the village of Rukshin in western Ukraine's Chernivtsi region. Local residents there complained of a mysterious vampire beast terrorising the community in recent months. They told how the beast had been feasting on their livestock – and had sucked all the blood from their bodies.

They said it had even managed to break into cages containing chickens and rabbits before massacring the animals – and was capable of leaping over 6ft-high fences.

The picture, however, is undoubtedly of a canid, although our resident veterinary consultant Shoshannah would not commit to whether it was a mangy feral dog, or a mangy fox, only that it was not like any bloodsucking phantasm of the Puerto Rican highlands that she had ever examined.

SOURCE: http://tinyurl.com/gmht7ea

Is it the Legendary Len Liggins?

Man Beasts (BHM)

Rocky Photo

The legend of "Sasquatch" or "Bigfoot" has been known far and wide for centuries. In Northern Colorado, a newly released photo snapped by a local couple claims to have captured evidence of "Bigfoot" lurking in the Rocky Mountains. "People from around the world are paying attention to this," said Michael Johnson, with Sasquatch Investigations of the Rockies.

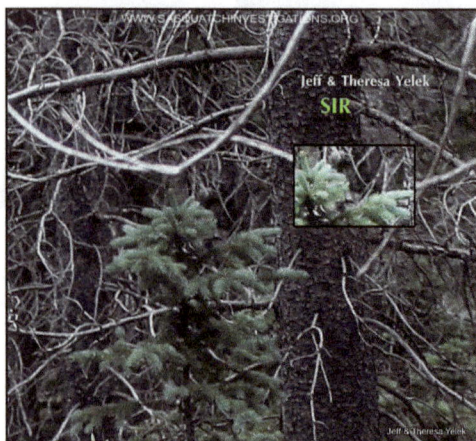

Johnson owns a museum in Denver dedicated to "Bigfoot." In it, he has a copy of the photograph shot by Jeff and Theresa Yelek. "If you focus [on it], you can see the nose, the black eye, the hair," he said.

The photograph was taken near North Park. We, I am afraid can see nothing even slightly bigfoot-like.

SOURCE: http://tinyurl.com/zn22x53

Beaver Patrol

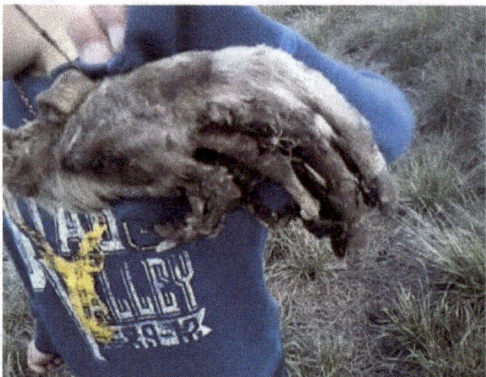

This photograph of this rather noisome object was doing the rounds earlier in the year. Was it a bigfoot hand? Well the fact that it was being held close to the camera does suggest that the picture is utilising forced perspective to make it look bigger than it is.

Various commentators on the *Bigfoot Evidence* blog have suggested that it is the paw of a beaver. I will admit that I was sceptical until I found *this* picture on Etsy.

Case closed I think.

SOURCE: http://tinyurl.com/jalx24r

Everybody must get stoned

I don't know what they have put into the water supply in Utah recently (you should also check out the Newsfile Extra section and you will see what I mean) but if anyone finds out what it is, can they please buy me a few grams?

Herewith is the most nonsensical crypto story of the year so far.

While on a hike near his home in Ogden, Utah, Todd May felt himself drawn toward something. "I would go out there often and find things, fossils, rocks. I looked around for about half an hour, then I saw it." Living in a hot spot for Bigfoot sightings, May said he had been interested in the mythological creature all his life.

He claimed a number of sightings and says that he would visit a hot springs in the area and often felt someone — or something — pegging him with rocks. May spotted what appeared to be a handprint on top of a rounded surface. He dug the large object out of the surrounding dirt and saw a familiar face. "It had the same facial structure as the creatures I had seen," he said. Since finding the object in 2013, May said he's had hundreds of people weigh in on their opinion about his finding. "There's haters out there, other Bigfoot enthusiasts that don't like that I found something first," May said.

People have noted the opinion of a Utah professor when the story first appeared who said the object was just a rock. "But that professor just saw the picture that was in the paper, he never saw it in person. When you actually see it, you can't help but see that it's a face," May said.

I would just like to point out that people who can correctly recognise rocks are not 'haters', and I also wonder whether he knows what 'pegging' means in the current British BDSM vernacular.

SOURCE: http://tinyurl.com/j7do7yc

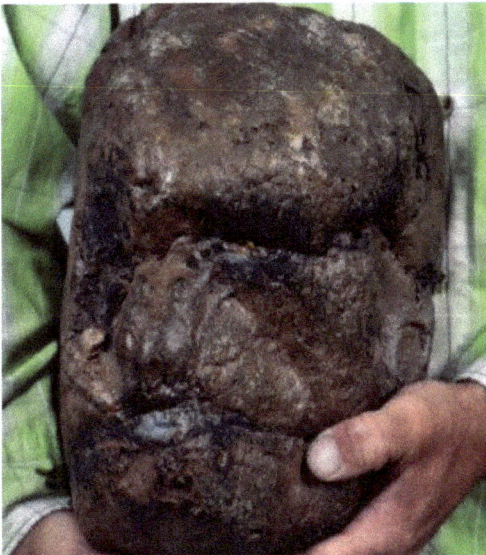

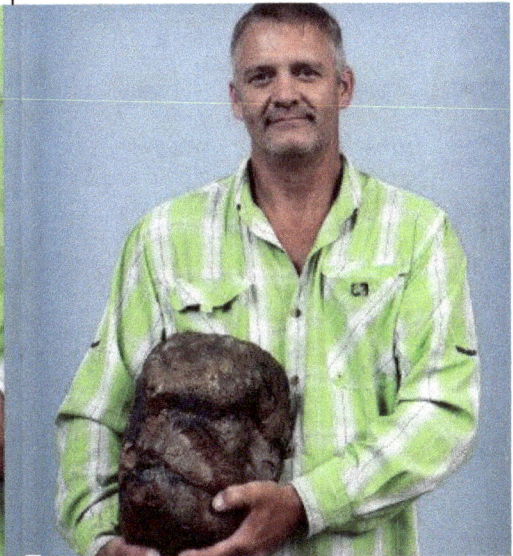

Mystery Cats

UK Round Up

- ### DEVONSHIRE

On 13th May, an East Devon newspaper claimed that: "A report of a big cat prowling the outskirts of West Hill has triggered dozens of people to come forward claiming they have spotted the creature around the Sid Valley and East Devon."

The story that triggered all this came when Jane Overthrow posted a status on the Ottery Matters Facebook page, saying she had seen a 'black panther-type' creature strolling across a field at the end of Lower Broad Oak Road, near Higher Metcombe, on Saturday morning.

She said the dark chocolate brown creature was heading towards a cattle field. "I honestly cannot tell you what it was, except it seemed too long and big to be a normal dog," said Jane. "We stopped to have a better look, but it was about 200 yards away."

Dozens of residents responded to the post to share their big cat encounters.

SOURCE: http://tinyurl.com/hfzyldu

- ### WORCESTERSHIRE

A lot has been happening in Worcestershire over the past few months. Dealing with the least impressive first:

I sent this picture to Max the other day claiming that it knocked his work on the lynx shot near Newton Abbot at the beginning of the 20th century into a veritable cocked hat. A dedicated panther spotter claims this is a picture of the beast prowling near Croome Court. Granted, it is hard to make out what

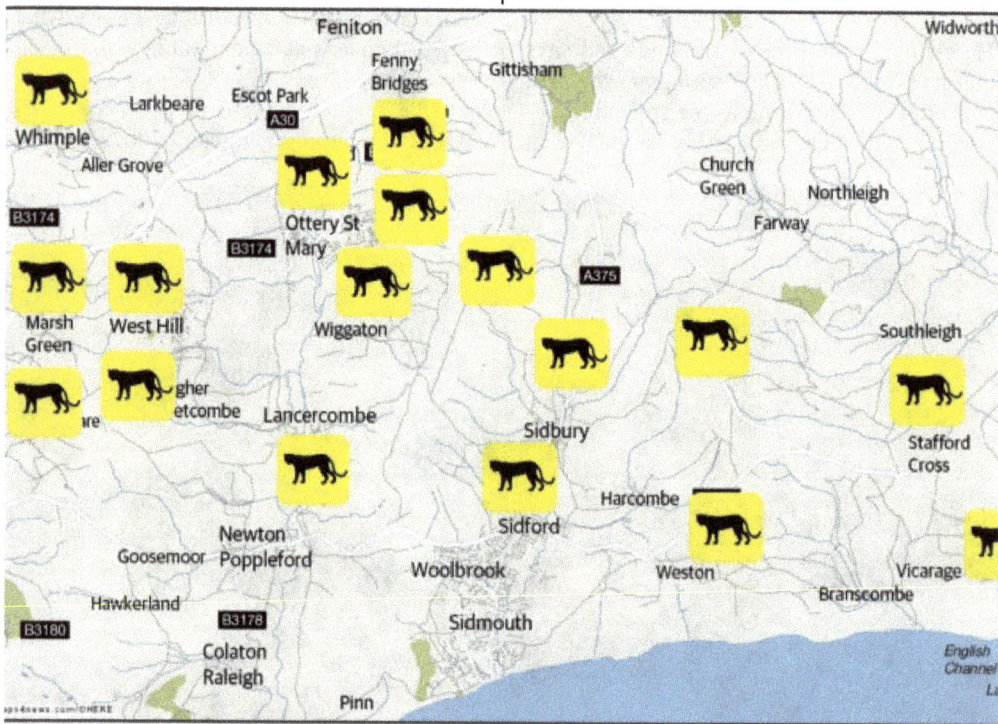

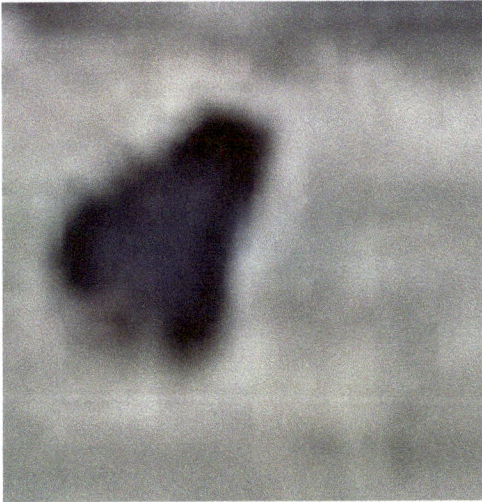

is exactly in the picture - but Paul Kear, from Worcester, is adamant he has seen the large black panther five or six times.

SOURCE: http://tinyurl.com/hndpob2

Then the *Daily Telegraph* reported that a couple - Robert and Nicola Ingram - have had an encounter with what they are convinced was a black panther while driving through Croome Court - a National Trust property - at 1am in early April. The couple said the creature weighed about nine stone and was as tall as their car window. They were convinced the big cat was going to pounce on their car before they sped off.

However the terrified couple then drew a sketch of a fearsome beast that they have dubbed the 'Werewolf of Worcester' because the drawing, although it is said to be a big cat, has features of a werewolf.

SOURCE: http://tinyurl.com/jr69bto

Aquatic Monsters

Oh No Pogo

The chief of the Okanagan Indian Band is accusing Dorothy Hawes (on the left of the picture with collaborator Maggie Parr) of cultural misappropriation. Hawes, a Victoria teacher and author, met with a representative from the Okanagan Indian Band before her book *Ogopogo Odyssey* was published.

"[The book] misappropriates our culture and our beliefs and our structures," said Chief Byron Lewis in an interview with CBC News. The book *Ogopogo Odyssey* tells the story of a young boy who has a chance sighting of the creature while visiting his grandparents in the Okanagan Valley, and meets a First Nations woman who tells him an indigenous story about the N'ha-a-itk, as it is referred to by First Nations.

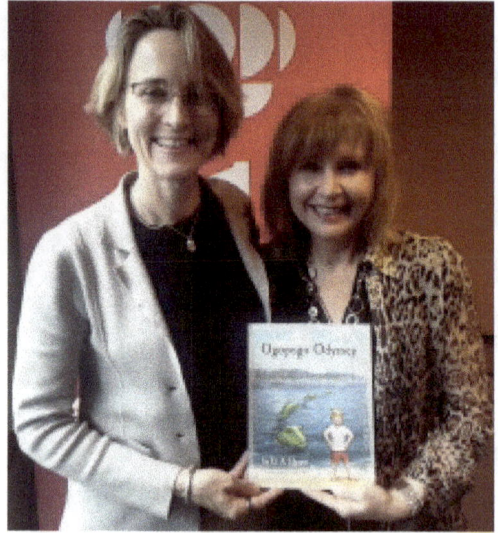

Chief Byron Louis confirmed that Hawes met with a person from the band's Territorial Stewardship Division in February 2016, but he said Hawes was advised during that meeting that the Okanagan Indian Band did not support her book and did not want her to publish it.

SOURCE: http://tinyurl.com/gsbodof

At least they didn't call it Whitey

A photograph showing a moving object in the middle of the Solent in late spring has caused eyewitnesses to question if they had seen a sea serpent. Three humps emerged from the water as the Nessie-shaped creature - first brought to the world's attention in 1933 -was thought to be making its way around the island.

The fascinating object was spotted in The Solent when Jo, from Newport, Isle of Wight, saw it move up and down when she was heading to Cornwall to see her family. She quickly took the photograph from the deck of Wightlink's Fishbourne to Portsmouth ferry.

The photo came only a week or so after a photographer with the unlikely name of Penn Plate claimed to have photographed a huge animal in the River Thames from the Emirates' airy cable car. Most people seem to have

© © YouTube / Penn Plate

written this particular incident off as a hoax, although others remember the unfortunate whale that swum into the river ten years ago. I would not be unduly surprised if either of these explanations turned out to be the truth.

SOURCE: http://tinyurl.com/hf5dg89

© Jo Wilde

Nessie News

© Facebook Help2Rehome Scotland

The best lake monster news of the past quarter comes—surprisingly—from Loch Ness. OK, don't get too excited. The picture above WAS FLASHED ACROSS the Internet in early July.

The gory scene was stumbled upon by a member of the public and provoked considerable debate on social media about the fate of Nessie. It emerged however, that the "body" will appear in the upcoming drama *Loch Ness*, with a film crew having been shooting scenes in the area. It included an evening and overnight shoot on Tuesday into Wednesday. It is understood that the carcass will feature in the show as part of a prank by local children to persuade people the monster had been found dead. The crew were filming primarily around the beach at the northern end of the loch. Few details have been released about the

programme, which will air early next year on ITV1.

SOURCE: http://tinyurl.com/zcdvme5

In mid April the 10-metre (30ft) model, from the 1970 film *The Private Life of Sherlock Holmes*, was discovered 180 metres down on the loch bed by an underwater robot, VisitScotland revealed. "We have found a monster, but not the one many people might have expected," Loch Ness expert Adrian Shine told the BBC News Scotland website. It would make a great exhibit for the Visitor Centre, and so I wonder how feasible it would be to emulate the Clive Cussler book *Raise the Titanic*?

SOURCE: http://tinyurl.com/zb7rzx2

Mexican carcass

A decomposed carcass found on a Mexican island in June, has left experts baffled as they try to discover what it is. The body of the creature was said to be "in such a bad state of decomposition, experts could not determine what it was," according to local Mexican media outlet BCS Noticias. It was found on the rocks of Cerralvo Island near the Mexican coast of Baja California Sur. Staff at the local Whale and Marine Sciences Museum near the site said the creature was most likely a beaked whale – however, these are allegedly not usually found in the Gulf of California.

However, people around the world, claim the specimen could even be a 'humanoid fish' or some form of alien creature. I am sure that they are right and that the experts are part of a conspiracy to suppress the truth.

SOURCE: http://tinyurl.com/j5ezgqh

In recent years one has got used to pictures of

Google Monster

boat wakes seen on Google Earth images of Loch Ness and other allegedly monster-haunted lakes being hailed as sea monsters. But this new image is somewhat more interesting, although we would not go as far as to confidently state it was cryptozoological in nature. The image was taken by satellite close to Deception Island in the South Shetlands. It appears to be a fin sticking out of the water, is about 30 meters (100 feet) from end to end. Considering that only a part of the object could be seen, the creature, if it is indeed an animal, could be far bigger.

SOURCE: http://tinyurl.com/z4o9vq3

Newsfile Xtra

Something that I have always found particularly irksome is that whereas our expeditions and research gets minimal attention from the world's mass media, a succession of stupid claims, and downright hoaxes are repeatedly splashed across the world's newspapers, websites and television screens.

Some years ago, for example, a father and son from Northern Ireland claimed to have conclusive proof in the existence of an enormous snake in Peru. We actually believe in such things, but the photographs the pair produced were palpably of a mud bank. We were polite about their claims but basically refuted them, whereupon not only did the Irish pair get TV and newspaper attention by the shedload, but - allegedly at least - they were offered large sums of money for their researches and ….. wait for it ….. we were repeatedly accused of not supporting their claims ONLY because they were not members of the CFZ.

Pah!

The latest bit of nonsense to fly around the world started out on June 2, when a video titled, "Possible bigfoot in Idaho!! Flying the drone around and ran across this" was posted on YouTube.

The duration of the 2-minute, 24-second features footage from a drone flying over an

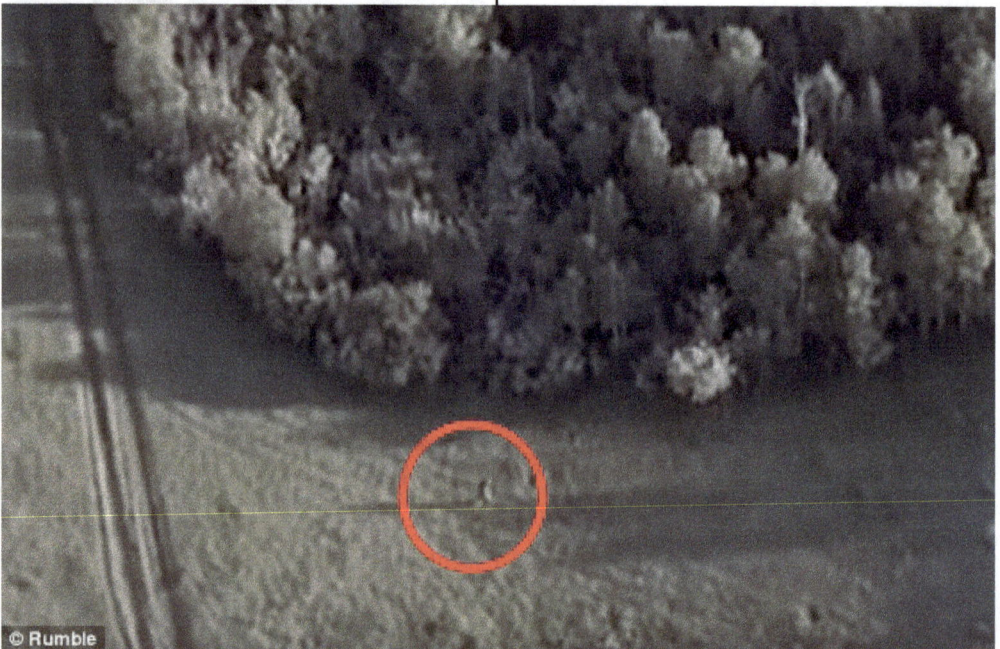

© Rumble

isolated area near Hawkins Reservoir about 35 miles south of Pocatello. As the drone passes over a grassy field, a mysterious figure runs toward a wooded area and hides. Over the course of about a week, the video went viral. It received more than 500,000 views and attracted attention across the world, being picked up by a variety of international news outlets.

Then, a couple of weeks later, the original uploader of the video, named hardpack101 on YouTube, posted two more videos revealing that it was all just a big joke.

The first of these newer videos depicts a man in a monkey suit running across the same field depicted in the original footage. As he approached the wooded area in the first video, he encounters another man dressed as a giant pink foot.

As the two men hug and goof off for the camera, the ominous music changes to an upbeat tone and one of the men holds up a sign that reads, "Bigfoot, really?"

In the second video the men goof off even further in the field.

The man who made and posted the videos is a longtime Pocatello resident who claims the videos were just a funny project made by him and his friend.

"We wanted to make a cool, believable video," the Pocatello resident said. "Some of the videos on YouTube were dumb as heck, so we thought, 'Hey, let's make a new one.'"

They thought at first the video could get a couple hundred of views. Instead, they got more than 500,000 views over the course of a week.

The reaction to the revelation that it was all a joke was predictable. After the hoax was revealed, the videographer said he received numerous threats from those who are strong believers in Bigfoot. The response has been so negative that the videographer wishes to

remain unidentified out of fear for his own safety. Back in 2003 after I dared to say that the 'thing' (as Ivan T Sanderson would have undoubtedly described it) that people (including me) saw in Bolam Woods, Northumberland could not be a flesh and blood giant ape, I received death threats and was kicked out of three bigfoot discussion groups hosted by Yahoo. So I am not at all surprised at what transpired to the anonymous videographer.

Don't get me wrong. I like jokes. I have played more than a few myself, but it is the way that the whole thing has so quickly got out of hand that I find disturbing.

Shortly after the original video went viral, a Bigfoot investigator from Utah even visited the site in Southeast Idaho where the footage was shot to obtain physical evidence. The investigator allegedly obtained footprint and hair samples from the scene and featured the investigation in a 40-plus minute YouTube video of his own.

When the videographer heard that the investigator was looking to have the hair specimens analyzed in a laboratory, he decided it was time to reveal the truth.

"It was getting a little out-of-control, and I wouldn't feel right having somebody spend thousands of dollars on lab tests to find out it was just a piece of fabric from my monkey suit." He later discussed the video with a number of Bigfoot investigators after it went viral. He was surprised because he said they failed to ask him one seemingly important question about the video.

"Not one person asked if I faked it," he said. "That's kinda weird. I think they just wanted to believe it was real so bad."

And therein lies the moral, or at least one of the morals, of the story.

Carl Marshall's Column

Humans seem to be heterogeneous in terms of the phenotypes we can potentially exhibit. Phenotypes, being the observable traits that are the result of our genetics and to a lesser extent our environment. Notwithstanding the obvious differences between peoples alive today, this variation constitutes to much less separation than what we have historically come to accept. What we call human in English initially refers to modern people, a term that would go on to also describe morality *humane* and intelligence *human-like*. But what we now consider 'human' is being challenged with new scientific discoveries along with the semantics (the study of meaning) that denotes the way we think.

So are we, the people of today, the only humans? Lets find out!

First of all, I want to briefly discuss the construct of race. The concept of race was originally created to justify separation and so called "superiority"; and throughout history there has been, and still are, individuals who misunderstand our phenotypes, and as a result believe that we are not all the same type of organism. Race in reality means very little and is simply a folk taxonomy that uses skin colour, culture, geography and language to place us into separate ethnic groups. Saying race does not exist is not ignoring the obvious differences we exhibit, it is acknowledging that this variation does not constitute to as much differentiation as we have historically believed. So for example, lets say you have a native Chinese and a native Argentinian, because genetic information is congenital from generation to generation and somewhat sporadic, yet at the same time relatively limited, you are extremely likely to find a Chinese and a Tanzanian that have more in common genetically with each other than they would with other individuals from their respected countries. When all this genetic material is taken into consideration you will likely find individuals that look nothing at all alike yet may have considerable genetic similarities.

The same can be said for any groups of humans alive today, whether it be Nepalese and native Americans or Australian aboriginals and Europeans. Even though we often categorise people based on appearance, this often leads us to make false conclusions in terms of which humans have more genetic similarity to one another, simply because when all the available genetic material is taken into context, it says otherwise! This has actually led some to believe that the different looking people of today are different sub-species of one species; Caucasoid, Negroid, Australoid etc, all

Are We The Only Humans Beings?

supposedly being sub-species of *Homo sapiens.* Now I want to make it absolutely clear that this is completely inaccurate both in terms of science and semantics. However, all of us as a whole are a sub-species! You and I, no matter what our ethnic background, are both *Anatomically Modern Humans,* a phrase used to describe our specific sub-species, known in trinominal taxonomy as *Homo sapiens sapiens.*

This extension has become more and more apparent in recent times as we continually discover fossil remains that are not technically *Anatomically Modern Human* but nonetheless belong to the same exact species as us. We understand that they are *H. sapiens* due to their capabilities meeting the definition of being human, for we have the capability to interbreed and exchange genetic material producing fertile offspring; this is important to understand as simply because two different organisms can produce offspring with each other does not necessarily mean both those animals are sub-species of the same species! If their offspring cannot produce fertile offspring with one other it is a genetic dead end. *Anatomically Modern Humans* on the other hand were once capable of producing fertile offspring with *Homo heidelburgensis,* which is the last scientifically classified distinct organism that we directly evolved from. We could also produce fertile offspring with Neanderthals *Homo neanderthalensis* and Denisovans *Homo sp. Altai* both of which evolved from *H. heidelburgensis* just as we did. And lastly this list also includes our closest non *H.*

Neanderthal

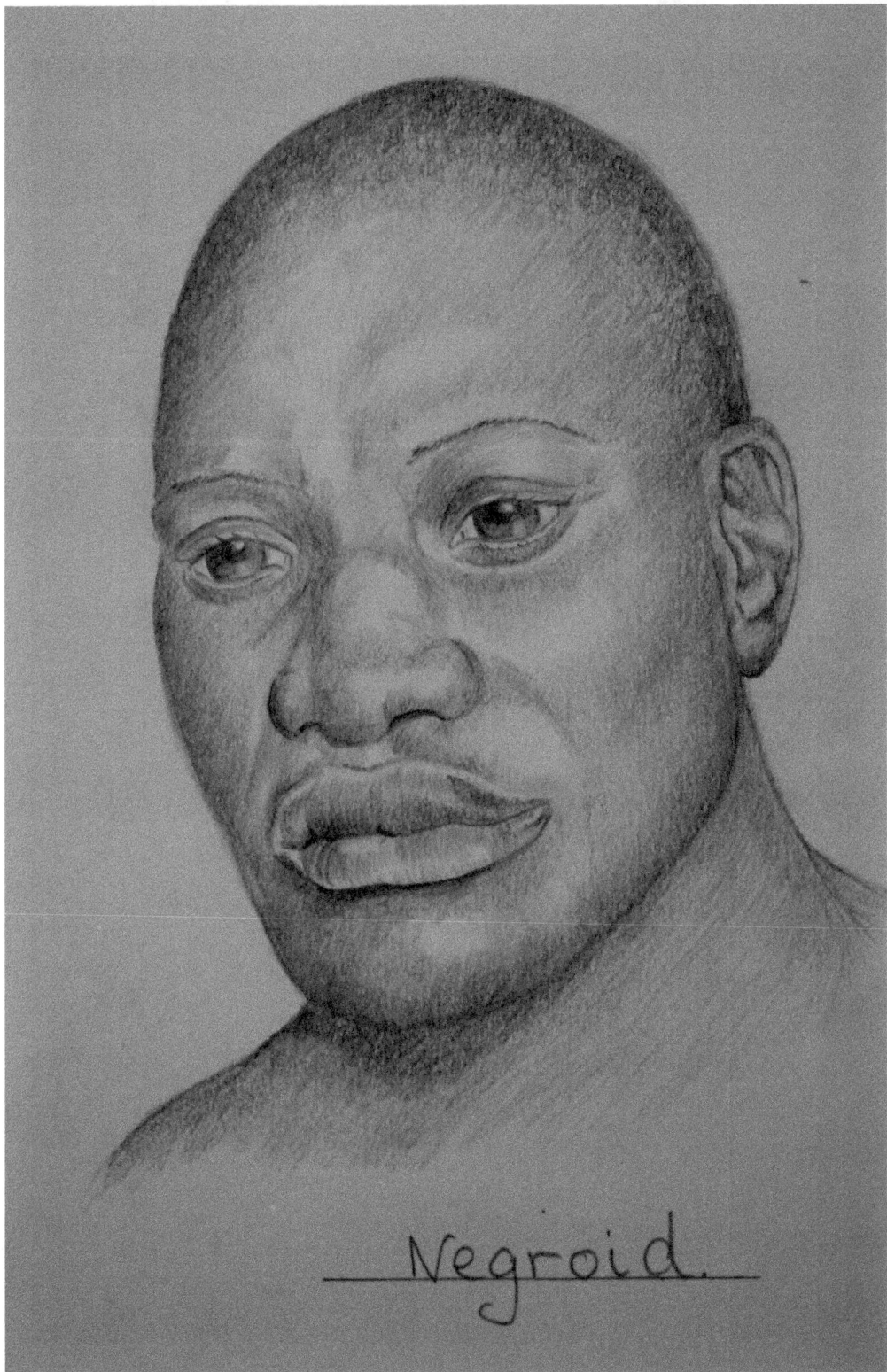

Negroid.

Australoid

Caucasoid

s. sapiens relative *Homo sapiens idaltu*. Strikingly similar to *Anatomically Modern Humans, H. s. idaltu* is not considered our sub-species due to anatomical differences and these distinct humans were most likely absorbed within our populations in Pleistocene Africa.

H.s. idaltu is an extinct hominan that most scientists feel comfortable with listing as a specific sub-species of *H. sapiens* along with us. Genetic and physical evidence is starting to make scientists rethink how to encompass our species, simply due to the scientific terminology specifying that these other organisms were just as human as we are! So not only were these other humans capable of interbreeding with our specific sub-species, but in terms of semantics they were the exact same species as us! This has been difficult to finalise though simply based on terminology and not capability. There have even been multiple hybrid specimens found which appear to be a mixture of *Anatomically Modern Humans* and *H. s. neanderthalensis* - modern humans with archaic features that are usually associated with Neanderthals!

Even today, with the means of genetic sequencing, we have been able to determine that all modern humans found outside of sub-Saharan Africa can have upwards of five percent of their DNA originating from Neanderthals or Denisovans. Denisovan DNA is unique, with about 3 - 5 percent of the DNA of Melanesians and Aboriginal Australians deriving from Denisovan bloodlines. The fact that this DNA exists within a majority of human populations today further shows us that our ancestors did interbreed, and that they were in fact capable of producing fertile offspring that could pass these genes on. One of the problems that makes this notion so unnecessarily controversial is the plain fact

that most people do not like the idea of their ancestors having interbred with these other hominans such as Neanderthals, even though they perfectly meet the requirements for them to be considered the same species as us. However, most people do not like this idea and its simply because we have for a long time presented Neanderthals as slow, dim-witted and immoral creatures, thus it would seem to be an insult to assume they have any genetic relationship with us at all, as all of us humans are pure; though recent DNA evidence paints a much more accurate picture!

Neanderthals, and one might reasonably assume Denisovans, displayed many behavioural traits that we associate with being human today. They buried their dead and sometimes even buried them with flowers. They cared for their sick and injured. They would talk to one another and had complex social cultures similar to how we do today. They even played musical instruments, had art and symbolism; these were beings that shared some of the most fundamental things that we associate with being human, even from a philosophical perspective.

So yes, me being human, with the majority of my ancestry being European in origin I have confirmed Neanderthal ancestry. I even display a fairly prominent brow arch, but that does not mean that I'm stupid, nor does it necessarily mean this particular feature derives from Neanderthals! Conversely, from a behavioural perspective some people believe the first humans were not us at all but the organism known as *Homo erectus* as many of the behaviours we associate with being human actually find their genesis with this species!

So are we, *Anatomically Modern Humans, Homo sapiens sapiens*, the only living humans? Well yes, at present, with the absence of confirmed novel hominids it would seem we are all that is extant, but from a genetic perspective absolutely not! This does, at least to some extent, lend credence to the long standing theory in cryptozoology that undiscovered hominids might still exist in some of the world's more remote regions, as from a purely biological perspective they should do, even if today it is unfortunately most likely only within our own genome. However, it is by no means impossible that in some truly isolated region of the world, small undiscovered populations of humans could potentially exist that have more in common with archaic types and also share higher genetic similarity! If discovered, these people would most likely only be found in the world's last extremely isolated regions such as the Taiga forest of Siberia, the Himalaya mountain range, the extreme northwest of the Americas, and also possibly Papua New Guinea where genetic dilution may have been more limited. Another reasonable hypothesis could be a closely related yet distinct species that does not exactly mirror our genetic characteristics and may fall outside the definition of our species that may have survived and evolved in isolation into modern times. The better we understand, the more we learn!

This brings me to my last and one of my most important points. We as a species try to group understandings using our language, but sometimes our old words do not correlate with our new understandings, and its up to us to correct this! This is because words express and shape how we perceive the world around us and how we process things, and yes, some people do not like this type of change because it will force them to re-evaluate their ideologies, but if words with old understandings are not fixed to meet the standard of our new understandings, we will only ever have misunderstandings!

Definitions are not supposed to shape the truth, truth is meant to shape our definitions!

Suggested Further Reading

- **Cox, P. Brian and Cohen, A. (2014).** *Human Universe.* HarperCollins Publishers.
- **Finlayson, C. (2009).** *The Humans Who Went Extinct* - *Why Neanderthals Died Out and We Survived.* Oxford University Press.
- **Harris, E. Eugene. (2015).** *Ancestors in Our Genome* - *The New Science of Human Evolution.* Oxford University Press.
- **Henrich, J. (2016).** *The Secrets Of Our Success* - *How Culture is Driving Human Evolution Domesticating Our Species and Making Us Smarter.* Princeton University Press.
- **Leakey, R. (1994). The Origin Of Human Kind.** BasicBooks Publishers.
- **Roberts, A. (2011).** *Evolution The Human Story.* Dorling Kindersley Publishers.
- **Stringer, Chris. (2012).** *Lone Survivors* - *How We Came To Be The Only Humans.* Times Books.

Carl Marshall works at Stratford Butterfly Farm and is a fine field naturalist. Over the past couple of years he has become a very enthusiastic member of the CFZ, and his quasi-Fortean view of British natural history fits in perfectly with my own. He was, therefore, the perfect choice as a columnist for the brave new *Animals & Men,* and we are proud to have him aboard.

World's first successful artificial insemination of southern rockhopper penguin

DNA tests have confirmed that one of the three southern rockhopper penguin (*Eudyptes chrysocome*) chicks born at Osaka Aquarium Kaiyukan between June 4 and 6 was conceived through artificial insemination. This is the result of a project led by Kaiyukan with the collaboration of Associate Professor KUSUNOKI Hiroshi (Kobe University Graduate School of Agricultural Science). It is the world's first successful case of a southern rockhopper penguin being conceived through artificial insemination. This species inhabits the southern islands near Antarctica such as the Falklands and they are listed as threatened on the IUCN Red List.

Kaiyukan and Associate Professor Kusunoki began their joint research in 2011, aiming to clarify the breeding habits of these penguins and to develop the expertise for their artificial insemination. In spring 2015 the group obtained a fertilized egg, but the chick did not hatch, and DNA tests determined that the unborn chick was the result of natural reproduction.

On this occasion more than one penguin was selected for breeding, and the group enlisted the cooperation of Tokyo Sea Life Park, where scientists had previously succeeded in

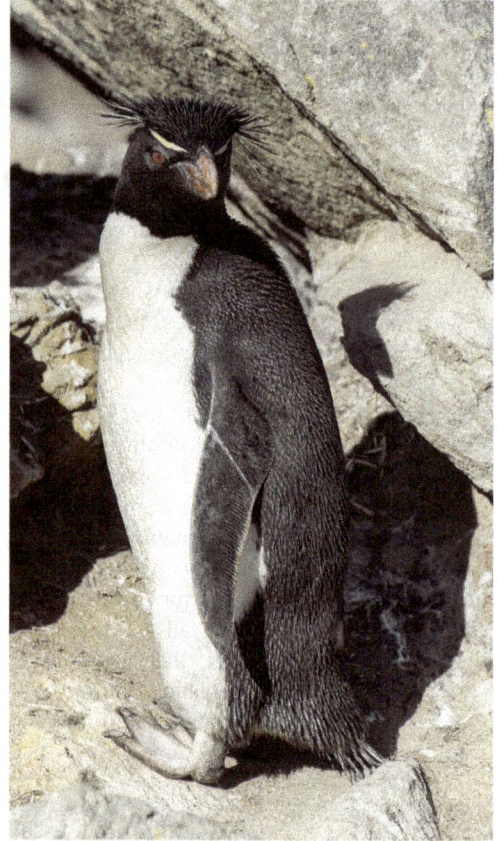

breeding this species through natural reproduction. At the end of April they obtained a healthy sperm sample from a male penguin at Tokyo Sea Life Park and transported it to Kaiyukan without loss of quality. At Kaiyukan they used blood tests to

Successful Captive Breeding

estimate the laying days of three female penguins and determine the best timing for artificial insemination. Between April 28 and May 4 the three female penguins laid five eggs between them. These were incubated by the penguin couples for approximately one month, and three chicks hatched between June 4 and 6. Results of DNA tests carried out on blood samples taken from inside the eggshells revealed that one of these chicks was conceived through artificial insemination.

SOURCE: https://www.sciencedaily.com/releases/2016/06/160628072251.htm

Rare spoon-billed sandpipers lay for first time in captivity

It was announced on 14th June that two female spoon-billed sandpipers (*Calidris pygmae*) had laid eggs in captivity for the first time. The spoon-billed sandpiper is one of the world's rarest birds, and the two females laid seven eggs at the Wildfowl and Wetlands Trust (WWT), Slimbridge, Gloucestershire. There are only about 200 breeding pairs of the critically endangered species left in the wild.

Nigel Jarrett, WWT head of conservation, said when staff discovered the first egg last week they "almost couldn't believe it". He said staff had "done their best" to enhance breeding conditions, with special lightbulbs, timer switches and lots of sand and netting to recreate the experience of migrating from tropical Asia to Arctic Russia. "For the last two years - ever since all the spoonies came into maturity - we've been doing everything to get these birds in the mood for love," he said. "And for two years we've come up scratching our heads and feeling a bit deflated. Now, we've had two mums busy laying and the significance of it is only just starting to hit home," he added. The WWT began trying to establish a flock at Slimbridge in 2011, as a back-up to the wild population which was declining by up to 25% a year. However, its lifestyle, which includes making an annual 10,000-mile round-trip between Russian Arctic breeding grounds and wintering grounds in South East Asia has meant that the bird has never been bred in captivity. The birds have been seriously affected by loss of habitat in East Asia and bird trapping by villagers in Bangladesh and Burma.

Mr Jarrett said the trust was now on the road to breeding spoon-billed sandpipers in captivity, which was "the ultimate insurance policy for the species in the wild".

SOURCE: http://www.bbc.co.uk/news/uk-england-gloucestershire-36524786

New and Rediscovered Birds

Spix's Macaw reappears in Brazil
24 Jun 2016

The last known wild Spix's macaw (*Cyanopsitta spixii*) - a large blue-feathered species of parrot that once lived in the rainforests of Brazil - disappeared at the end of 2000, and there are thought to be fewer than 100 surviving in captivity.

Primarily as a result of trapping for trade as well as loss of habitat, this critically

ARA SPIXI.

endangered macaw was thus thought possibly extinct in the wild, although 130 remain as part of a captive breeding programme. On 18th June, a bird was sighted by a local farmer and then filmed the next day.

The bird was first sighted on 18th June by local farmer Nauto Sergio de Oliveira. On the following day, his neighbour Lourdes Oliveira and daughter Damilys woke up before dawn to look for the macaw in Barra Grande creek's riparian forest. At 6:20 AM they found and filmed it.

Biologists from the Society for the Conservation of Birds in Brazil (SAVE Brasil, BirdLife Partner), one of the organisations that make up Projeto Ararinha na Natureza (Spix's Macaw in the Wild Project) which aims to bring the bird back from extinction, were contacted and the distinctive vocal calls confirmed that it was indeed a Spix's macaw. Pedro Develey, SAVE Brasil's Director, immediately told other project members and organised an emergency trip to Curaçá to locate the bird. He said, "The local people were euphoric. They set up a WhatsApp group to coordinate and maximise the search for the bird, and ensured no potential dealers could enter the area."

The bird's origin is uncertain, but it is quite possible that it was released from captivity. Conservationists have had a large presence in the area where it would likely have been seen, and recent patrols and project warning signs against trapping might have prompted a panic release.

However, this macaw can live for 20-30 years in the wild (more in captivity) and the area is very large with access to some parts being difficult. Develey added, "We don't know yet, and that makes it even more interesting." He also said, "As far as I know there is no missing bird in the breeding centre in Brazil, that is located more than 2000km from the site," adding weight to the idea the bird has recently been released by a private collector.

Desperate efforts have been made in recent years to save Spix's macaw from extinction, including captive breeding efforts in Brazil, and in Germany and Qatar.

Develey continued, "Now we have a model to understand the bird's behaviour in the wild, we can understand what to do when we release the captive birds in Curaçá."

SOURCE: http://www.birdlife.org/americas/news/spix%E2%80%99s-macaw-reappears-brazil

Extremely rare 'Species X' rediscovered in Brazil after 75 year disappearance

In May it was announced that the blue-eyed ground-dove (_Columbina cyanopis_) had been rediscovered in Brazil. It had not been seen since 1941. The site of its recent location in Minas Gerais, in the Brazilian state is being kept top-secret, but researchers can only confirm sightings of 12 individuals, so securing its habitat will be the key to conserving this elusive bird, previously known from a handful of stuffed and ageing museum specimens.

"When he played the video there was a commotion in the crowd and non-stop applause," said Pedro Develey, SAVE Brasil (BirdLife in Brazil). "It was pure emotion."

The group of researchers - supported by SAVE Brasil, Rainforest Trust, and Butantan Bird Observatory – had been working in secret to scientifically report the rediscovery, and to attempt to develop a conservation plan that secures the critically endangered bird's long-term survival. Rafael Bessa said, "I returned to the place and I could recreate this vocalization with my microphone. I reproduced the sound and the bird landed on a flowering bush, coming towards me. I photographed the animal, and when I looked at the picture carefully, I saw that I had recorded something unusual. My legs started shaking."

The blue-eyed ground-dove occurs exclusively in Brazil and is threatened by the destruction of the Brazilian Cerrado, a savannah-like habitat. "We are now worried about the conservation of the species," explained Bessa. "We are working on several fronts to build this plan. The main action is to ensure that the area where it was found becomes a protected area, which would benefit not only the blue-eyed ground-dove, but many other threatened species occurring there."

SOURCE: http://www.birdlife.org/americas/news/extremely-rare-species-x-rediscovered-brazil-after-75-year-disappearance

Endangered species of bird comes into view after 178 yrs

Also in May, a rare and endangered species of bird which was first spotted in Chitwan around 178 years ago, and not seen again, has been rediscovered. A red-faced liocichila (*Liocichila phoenicea*) - known as 'Simrikane Lito Shila' in Nepali - was discovered at Chisapani in Dahakhani by a

team of ornithologists from the Bird Education Society and the Nepalese Ornithologists Union (NOS). "With this rediscovery Chitwan now boasts of being a home to 631 species of birds," said former President Bidari.

According to data, a total of 878 species of birds are now recorded in Nepal.

The team comprised ornithologists Dr Hem Sagar Baral, and NOS Eastern Region Coordinator Badri Chaudhary among others.

SOURCE: http://kathmandupost.ekantipur.com/news/2016-05-23/endangered-species-of-bird-comes-into-view-after-178-yrs.html

Fossils

Fossil pelagornithid bird with 6m wingspan found in Antarctica

It was reported in May that the fossil remains of a 50-million-year-old bird with a six-metre wingspan had been found in Antarctica.

Paleontologists at a natural history museum in Argentina said they had identified the pelagornithid, or bony-toothed bird, nearly three years after its fossilised bones were first found at an Argentine research base on the

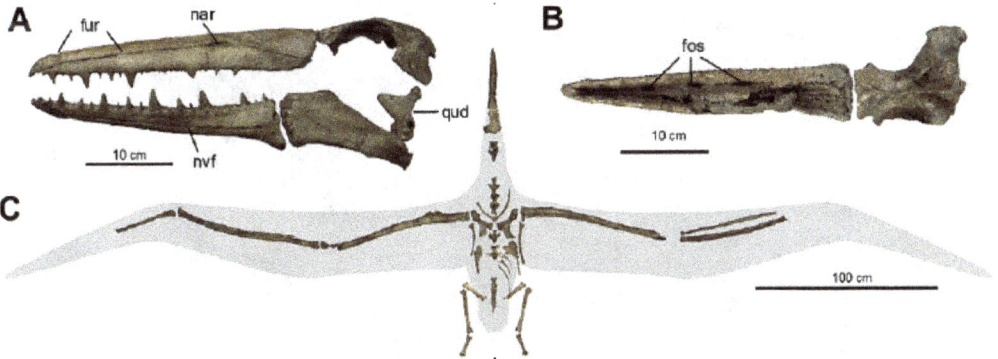

Antarctic island of Marambio.

"Almost three years ago, remains began to appear of what we believed could be this bird. Then we found a bone that confirmed that it was a pelagornithid, an extinct family of enormous seabirds," said Carolina Acosta Hospitaleche, a researcher on the project.

The bird's wings, fully extended, spanned more than 6.4 metres, which is twice as big as the albatross, the largest flying bird alive today, which has a wingspan of 3.3 metres.

Ms Acosta Hospitaleche's colleague Marcos Cenizo, the director of the Natural Sciences Museum of La Pampa, said the bird was the largest pelagornithid specimen ever found. "The shape of their wings allowed them to glide and cross large distances across the oceans," he said.

According to Antarctica specialists, there were two kinds of pelagornithid on the continent, one that reached up to five metres tall, with a similar wingspan, and another that stood more than seven metres.

The researchers said, "The birds likely developed to their monstrous size some 50 million years ago, when warming ocean temperatures would have given them an abundance of food to thrive on."

But the recently identified specimen would have been quite light despite its stature, at 30 to 35 kilograms —"almost like a feather," Mr Cenizo said. The researchers published the find in the *Journal of Paleontology*.

SOURCE: http://www.abc.net.au/news/2016-05-19/giant-prehistoric-bird-found-in-antarctica/7427460

Early bird wings preserved in Burmese amber

An international research team, led by Dr Xing Lida from the China University of Geosciences, and colleagues from Canada, United States and Professor Mike Benton from the University of Bristol, UK, have uncovered fossil wings in an amber deposit (solidified tree sap; the Burmese amber occurs in small blocks that are polished to reveal the treasures within) in northeastern Myanmar, which has produced thousands of exquisite specimens of insects of all shapes and sizes, as well as spiders, scorpions, lizards, and isolated feathers. This is the first time that whole portions of birds have been noted.

The fossil wings are tiny, only two or three centimetres long, and they contain the bones of the wing, including three long fingers armed

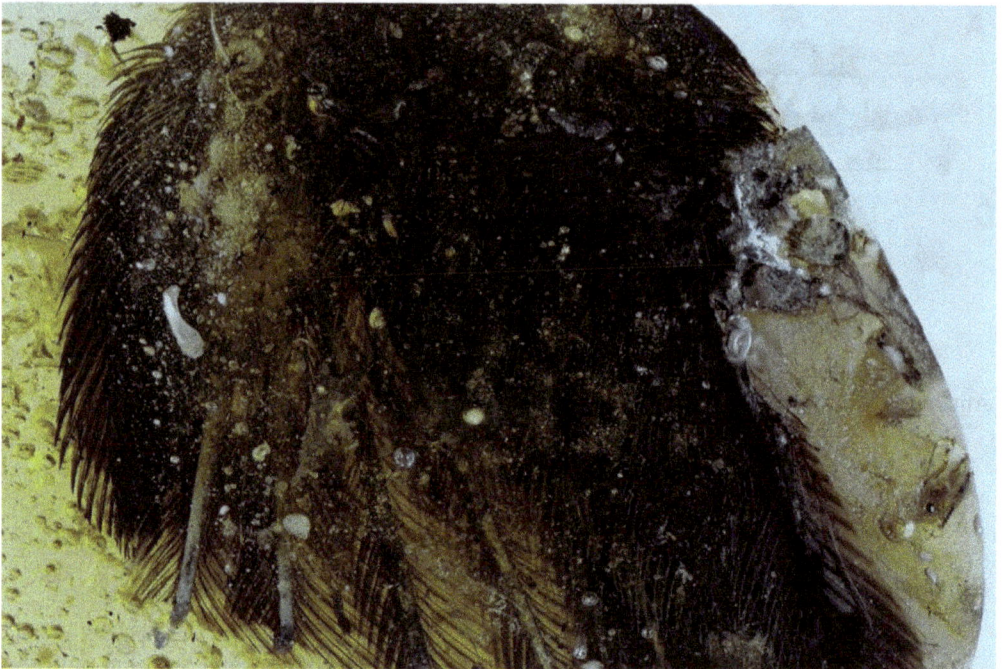

with sharp claws, for clambering about in trees, as well as the feathers, all preserved in exquisite detail. The anatomy of the hand shows these come from enantiornithine birds, a major group in the Cretaceous, but which died out at the same time as the dinosaurs, 66 million years ago. Mike Benton, Professor of Vertebrate Palaeontology from the School of Earth Sciences at the University of Bristol and one of the researchers, said: "These fossil wings show amazing detail. The individual feathers show every filament and whisker, whether they are flight feathers or down feathers, and there are even traces of colour -- spots and stripes."

Dr Xing Lida, lead author of the study, explained: "The fact that the tiny birds were clambering about in the trees suggests that they had advanced development, meaning they were ready for action as soon as they hatched. These birds did not hang about in the nest waiting to be fed, but set off looking for food, and sadly died perhaps because of their small size and lack of experience. Isolated feathers in other amber samples show that adult birds might have avoided the sticky sap, or pulled themselves free."

The research is published in *Nature Communications*.

Journal Reference:
Lida Xing, Ryan C. McKellar, Min Wang, Ming Bai, Jingmai K. O'Connor, Michael J. Benton, Jianping Zhang, Yan Wang, Kuowei Tseng, Martin G. Lockley, Gang Li, Weiwei Zhang, Xing Xu. Mummified precocial bird wings in mid-Cretaceous Burmese amber. Nature Communications, 2016; 7: 12089 DOI: 10.1038/ncomms12089

SOURCE: https://www.sciencedaily.com/releases/2016/06/160628141419.htm

Rare and/or out of place bird sightings from around the world

Rare, unusual bird spotted in Seychelles by local university student

In June a student from the University of Seychelles spotted a greater painted-snipe (*Rostratula benghalensis*) – a first for the Seychelles - a group of 115 islands in the western Indian Ocean.

It was seen at Perseverance, a reclaimed island close to the centre of Mahe, the most populated island of the archipelago. "To spot a rarity I think it is up to chance, but if you have some special skills to spot elusive birds, then when you are on the field, you know what you are looking for," the student told SNA.

The species is a rarity among birds as the female is larger and has brighter coloured feathers than the male - usually in the bird family, it is the opposite. The species is characterised by its slightly downward curving bill used for probing prey in water and mud with long partly webbed feet adapted for walking in muddy areas.

"When I spotted the bird, I knew straight away that it was a snipe and one that I hadn't spotted before. So, with camera in hand, we became the paparazzi and it (the bird) the unique superstar," said Onezia.

Another rare sighting was of a Eurasian reed warbler, which landed on a boat off the coast of Alphonse during a fly-fishing trip.

Records of sightings of avian visitors in the Indian Ocean Islands are kept by the Seychelles Bird Records Committee (SBRC), set up in 1992 to confirm the identity of migratory and vagrant species in Seychelles.

To date, 272 species have been recorded in Seychelles, including 62 breeding species,

28 yearly migrants and 172 vagrants.

SOURCE: http://
www.seychellesnewsagency.com/
articles/5342/
Rare,+unusual+bird+spotted+in+Seychelle
s+by+local+university+student#sthash.2R
Nx9P6K.sP5L4bNs.dpuf

Endangered bird found in China

A rare bird species typically found in India and Southeast Asia was spotted in southwest China in June, providing valuable material for research of this endangered species.

A beautiful nuthatch (*Sitta formosa*) was pictured among a group of other bird species during a field inspection in Yunnan's Daweishan National Nature Reserve from April to May, Mo Mingzhong, a wildlife official in Honghe Hani and Yi Autonomous Prefecture, said on 30th May..

"The beautiful nuthatch is extremely rare and was barely found in China before, so it is quite significant to even have taken a picture of the bird," Mo said.

Chinese scientists obtained a specimen of the species for the first time in 1972, but there has been no systematic research since then, said Wu Fei, an ornithologist with the Chinese Academy of Sciences.

The beautiful nuthatch is listed as "vulnerable" on the IUCN Red List of Threatened Species.

SOURCE: http://news.xinhuanet.com/
english/2016-06/06/c_135416916.htm

Conservation

New Sociable Lapwing habitats discovered in Uzbekistan

The sociable lapwing (*Vanellus gregarious*) is one of the world's rarest and most threatened birds. It breeds in Kazakhstan and southern Russia and winters from Sudan to Pakistan and India. Hunting along its migration routes is considered one of its main threats, and conservationists are eager to learn of how it gets from its breeding grounds to its wintering areas. Very little is known about their path along the eastern flyway, from Kazakhstan to Pakistan and India. So when UzSPB (BirdLife affiliate in Uzbekistan) found 400 sociable lapwings at a reservoir in southwestern Uzbekistan in 2012, and when a few of the birds fitted with satellite tracking devices in Kazakhstan turned up in the same area and in adjacent parts of Turkmenistan, experts' interest was piqued. New research from last year shows that this area, known on both sides of the border as Tallymerdzhan – is used by possibly the species' entire eastern flyway population and perhaps a third or more of its global population.

In October 2015, researchers from Turkmenistan, Uzbekistan and the UK carried out coordinated surveys of the area in both countries, and found as many as 4.225 birds in Uzbekistan and 3.675 in Turkmenistan. The total number of birds using the area was estimated at between 6.000 and 8.000. Birds use this area for around two months while they fatten up for the crossing of the Hindu Kush mountains that takes them to their wintering grounds, one of the longest stopover periods ever recorded in a long-distance migrant.

The discovery of the large population of sociable lapwings in the area suggests that the eastern flyway is as important in terms of numbers as the much better studied western flyway (that goes from Kazakhstan south into Syria and Saudi Arabia), and that Tallymerdzhan is one of the most important sites for the species globally.

The eastern flyway population is substantially larger than previously known, and more work is required in range countries along this flyway to ensure that threats are monitored and minimised.

SOURCE: http://tinyurl.com/js6mnd4

The Bahama Oriole – threatened species

The Bahama oriole (*Icterus northropi*) is the rarest bird in the Bahamas - and one of the rarest in the world - with fewer than 260 remaining on a single island.

The oriole is the most threatened of the six

endemic bird species found in the Bahamas; it once lived on Abaco as well, but disappeared from that island in the 1990s.

"Previous studies indicated that the Oriole relied on coconut trees for nesting, and was in serious decline because these trees were dying from lethal yellowing," according to BNT Science & Policy Director Shelley Cant-Woodside. "The current research is key to understanding the biology of this amazing species. We now know that the Bahama Oriole is not completely reliant upon coconut trees, but uses other trees in residential areas and in the pine forest. This is an important find if we want to ensure a secure future for this very rare bird."

Research on the Oriole was last conducted in 2011. The current University of Maryland team is led by Dr Kevin Orland. Their work is funded by the American Bird Conservancy, a Virginia-based group that promotes bird conservation throughout the Western Hemisphere, and aims to survey populations of the bird on Andros; determine the number of nesting pairs in residential and forested areas; confirm whether coconut trees are the preferred nesting tree; and capture the basic biology of the species, including food sources and predators.

Bahamians are also being trained by the researchers in field research and promoting a wider appreciation of this rare bird, which is threatened by forest fires, logging, introduced diseases, invasive species, and the potential effects of climate change in terms of sea-level rise and changes in habitat.

"A lot of work still needs to be done," Ms Cant said, "but through partnerships like this, coupled with local stakeholder involvement, more reliable information will be obtained to help improve conservation measures."

Follow the progress of this project by joining the Bahama Oriole Project Facebook page.

SOURCE: http://tinyurl.com/zk5bydf

Call for North Pennine farmers to help save the Curlew

The RSPB, is calling on farmers in the North Pennines to give a home to nesting curlews (*Numenius arquata*). This wading bird could become extinct in a generation unless urgent action is taken. The UK is one of the most important countries in the world for breeding curlews, hosting up to a quarter of the global breeding population. But since the '90s, their numbers here have almost halved. But the good news is that in the North Pennines, farmers can take steps to help reverse this serious decline. RSPB Conservation Advisor, Janet Fairclough, provides specialist advice to farmers in the North Pennines about how they can help wildlife thrive alongside their agricultural businesses. She says: "Traditional hay meadows provide excellent habitat for nesting curlews. By shutting meadows up in the spring and putting off mowing them until July, farmers can give curlews enough time to nest and raise their chicks. If you need to cut meadows before July, keep an eye open for curlews flying up in front of the tractor, as they may have come off a nest. Mowing from the centre of the field outwards can help push any flightless chicks out of the way of machinery and into the safety of neighbouring fields."

"Farmers can also give curlews a home by maintaining rush pasture and allotments to provide a mixture of short and long vegetation across the farm. Grazing with both cattle and sheep provides the vegetation structure that curlews prefer. Rush management by cutting or weed-wiping may also be necessary to keep them from becoming too dense."

The dramatic decline of curlews in the last few decades has been caused by the low number of chicks fledging. This in turn, is the result of a loss of suitable breeding habitats due to agricultural intensification and increased predation from foxes and crows.

The RSPB has launched a five-year recovery programme which includes research on a series of trial sites across the UK, to test whether a combination of habitat management and predator control can be effective in halting the curlew decline across the wider landscape.

RSPB
20 Jun 2016

SOURCE: http://tinyurl.com/hz2ynqs

Other News

World's oldest known Terek Sandpiper discovered in Belarus

The Terek sandpiper (*Xenus cinereus*) is a rare species to be found in Belarus. During May ornithologists from the Academy of Sciences of Belarus at the birds ringing station in the Turau Meadow, Belarus were surprised to find one such bird with the band on its leg showing it was 17 years old, the oldest of its kind in the world, having completed 200.000 kilometres of flight.

They discovered that this particular bird had been banded as a chick in a meadow near the village of Zapesochye on 21 June, 1999 – the year the Turau Meadow ringing station was founded. Since then, 'meetings' have occurred between the bird and ornithologists during re-catching in 2005, 2011 and now in 2016.

Up until then, the known maximum age of a tagged Terek sandpiper was 16 years – on a bird found in Finland. The Important Bird and Biodiversity Area (IBA) of Turau Meadow seems to be a favourite spot for the seniors of the species: recently, two other Terek sandpipers – 14 and 15-year-old birds – were caught there, said Pavel Pinchuk, director of the Belarusian birds ringing centre.

In recent years, there have been more and more opinions that banding as a way of studying birds is becoming outdated and is no longer effective enough. Metal bands are being replaced by modern equipment, but a transmitter fixed to birds and transmitting signals will never stay alive as long as a simple band; the lifespan of a transmitter is usually only a few years and it can fall prey to technical issues. Finding a bird with a 17-year-old band still attached to it shows that this method's importance cannot be underestimated.

The fact that this old Terek sandpiper came back to the same spot more than once for the last 17 years also shows that it is vital for birds to have a safe site that they can return to, and the Turau Meadow IBA is home to thousands of waders and other wetland birds. APB (Birdlife in Belarus) is working to ensure it stays that way.

SOURCE: http://www.rarebirdalert.co.uk

Hen Harrier 'action' plan in ruins as another bird disappears and nests are down

The RSPB has announced the disappearance of yet another satellite-tagged hen harrier (*Circus cyaneus*); a female named Chance that was tagged in 2014 and whose last known location was on a South Lanarkshire grouse moor.

It comes just a few weeks after the news that another satellite tagged harrier, Highlander,

vanished over a grouse moor in Durham.

Here is the full statement on the RSPB website:

"It is with a heavy heart that only weeks after our beloved Highlander vanished over a moor in Durham, I have to share the news that our one remaining satellite-tagged hen harrier, Chance, has now also disappeared.

For those who haven't been following this blog, Chance was a female hen harrier, named by RSPB Scotland, who was tagged in June 2014 by members of the Scottish Raptor Study Group before the Hen Harrier LIFE+ Project began. However, the project followed her movements closely. RSPB staff who were monitoring Chance became concerned when her tag suddenly and inexplicably stopped transmitting at the end of May. A search of her last known location, on a South Lanarkshire grouse moor, was carried out by RSPB Investigations staff, but there was no sign of her. It is possible that she could have moved some distance from here before going offline. We don't know what caused the satellite tag to fail but as with Highlander, transmission up to that point had been strong and there was no indication of battery failure.

She has not been found.

Needless to say, we are deeply saddened, disappointed and frustrated at the disappearance of Chance. We were looking forward to following her movements, monitoring any nesting attempts, and sharing them on the LIFE+ project website. We had high hopes that now in her second year, this would be the summer she raised a brood of her own.

We appeal to anyone who can provide any information about Chance's disappearance to contact the RSPB in the first instance, or if the circumstances appear suspicious, Police Scotland on 101. You can also read a full statement on Chance's life on the Hen Harrier LIFE Project website."

At the same time as announcing the disappearance of yet another hen harrier on a grouse moor, the RSPB made public that a late nest has been found on their Geltsdale reserve in Cumbria. It is now known that this nest is just one of three active nests in England this year!

So that's three nests in the first breeding season since the publication of the Hen Harrier Joint Action Plan - half the number of successful nests in 2015. So despite DEFRA's 'action plan' 2016 is set to be another catastrophic year for hen harriers, with illegal persecution on grouse moors continuing unabated.

It is clear that landowners running driven grouse shoots are paying no attention to the Hen Harrier Joint Action Plan let alone the law, and are continuing to instruct their gamekeepers to carry on the killing of not just hen harriers, but any bird that are a threat to grouse.

27 June 2016

SOURCE: http://www.rarebirdalert.co.uk

Fairywrens learn calls before hatching

In June, it was announced that new research had demonstrated that some songbirds start learning to imitate their parents' vocalisations before they even hatch. In a study, all the

red-backed   fairywren   (*Malurus melanocephalus*) females concerned called to their eggs while incubating, and most continued to call to their nestlings for five to six days after they hatched; as a result, mother and offspring calls were more similar than would be expected by chance. Parents also put more effort into feeding nestlings with calls similar to their own. "Fairywrens have become a new model system in which to test dimensions in the ontogeny of parent-offspring communication in vertebrates," said Mark Hauber of New York City's Hunter College.

Diane   Colombelli-Négrel   and   Sonia Kleindorfer   of   Australia's   Flinders University, and their colleagues from Cornell University, along with Hauber, discovered that superb fairywren (*Malurus cyaneus*) nestlings learn to imitate their mothers' calls while still in the egg, so they wanted to see whether this behaviour extended to other species and to learn more about its ecological context. They turned to the related red-backed fairywren for the study.

"Prenatal vocal learning has rarely been described in any animal, with the exception of humans and Australian Superb Fairywrens," said Dr William Feeney of the University of Queensland, an expert on the interactions between cuckoos and host birds. "In this study, the authors present data suggesting that, like the Superb Fairywren, Red-backed Fairywrens also learn their begging calls from their mother. This result is exciting as it opens the door to investigating the taxonomic diversity of this ability, which could provide insights into why it evolves."

The original discovery was a fortuitous accident. "Because fairywrens have high predation rates, we originally placed microphones under Superb Fairywren nests to record alarm calls against predators twenty-four seven," says Colombelli-Négrel. "As a result, we discovered embryonic learning in Superb Fairywrens."

Reference
Colombelli-Negrel, D, Webster, M S, Dowling, J L, Hauber, M E, and Kleindorfer, S. 2016. Vocal imitation of mother's calls by begging Red-backed Fairywren nestlings increases parental provisioning. The Auk: Ornithological Advances 133: 273-285.

SOURCE: http://www.birdwatch.co.uk/channel/newsitem.asp?c=11&cate=__16284

World's smallest gull breeding in Scotland

for the first time

On 16th June, it was announced that, for the first time, the world's smallest species of gull, the little gull (*Hydrocoloeus minutus*) has been confirmed to be nesting in Scotland at RSPB Scotland's Loch of Strathbeg nature reserve. RSPB Scotland staff will mount a 24-hour watch and use cameras to protect

these rare birds. This is the first confirmed breeding record for this species in Scotland, and the sixth from across Britain since the 1970s at least. The most recent record was from Norfolk in 2007.

Richard Humpidge, RSPB Scotland Sites Manager, said: "We're really excited to have these smashing little birds nesting on the reserve. A few years back, we did a lot of work on our tern nesting island reshaping it and adding 10 tons of shingle and shelters as well as installing a fence around the edge of the pool to prevent access for ground predators. It's been a great success: four years ago there were just 10 pairs of common terns and they failed to raise any chicks, the next year there were 60 pairs and this year we have 130 pairs and their eggs are just starting to hatch. It's great that the Little Gulls are using the same area and we hope that it will also give them the protection they need to raise chicks when their eggs hatch shortly."

Little gulls are the smallest species of gull, weighing not much more than a blackbird.

RSPB Scotland

SOURCE: http://tinyurl.com/gqbemfb

New immigrant: Shiny cowbirds noted from a recording altitude of 2,800 m in Ecuador

At the beginning of May came the report that two young shiny cowbirds (*Molothrus bonariensis*) - a parasitic bird that lays its eggs in the nests of other birds – had been seen in the Andean city of Quito, Ecuador, for the first time. This, in turn, represents an altitudinal expansion of approximately 500 m.

The research team, led by Dr Verónica Crespo-Pérez, professor at Pontificia Universidad Católica del Ecuador (PUCE) think that the breeding populations might have been impelled by forest fragmentation and/or climate change. Now the 'immigrants' could be threatening native birds. The study is published in the open access *Biodiversity Data Journal*.

"The shiny Cowbird is native to the lowlands cof South America but within the last 100 years, it has been expanding its distribution to higher altitudes and latitudes" says the lead author. The bird had already been noted from high altitudes in Bolivia and Perú, and in some localities in the Ecuadorian Andes, and since 2000, Juan Manuel Carrión, co-author and director of the Zoo in Quito, recalls seeing shiny cowbirds near his home in a valley near Quito at 2,300 m above sea level (asl), but one has never before been reported from an altitude as high as 2,800m asl.

The observations took place at the PUCE campus about a year ago. Two juvenile shiny cowbirds were seen parasitizing two

different pairs of rufous-collared sparrow, one of the most common birds in Quito. The cowbirds displayed food-begging behaviours to adult sparrows, including chasing the sparrows on the ground and chanting intensely on bushes and tree branches. "These observations mean that the birth mother of the cowbird laid her eggs in the nests of the sparrows, who inadvertently, became the cowbird's foster parents and incubated, fed and cared for the it as if it were its own, even though the cowbird is almost twice as big," says Miguel Pinto, co-author and professor at Escuela Politécnica Nacional.

"The sparrows were not feeding fledglings of their own species, which suggests that the Cowbird could be having some negative effect on the Sparrow, at least on their ability to reproduce," points out Tjitte de Vries, co-author and professor at PUCE.

Journal Reference:
Verónica Crespo-Pérez, C. Miguel Pinto, Juan Manuel Carrión, Rubén D. Jarrín-E, Cristian Poveda, Tjitte de Vries.The Shiny Cowbird, *Molothrus bonariensis* (Gmelin, 1789) (Aves: Icteridae), at 2,800 m asl in Quito, Ecuador. Biodiversity Data Journal, 2016; 4: e8184 DOI:10.3897/BDJ.4.e8184

SOURCE: https://www.sciencedaily.com/releases/2016/05/160504121812.htm

For those of you not aware, as well as this column in *Animals & Men,* Corinna writes a daily Fortean bird blog which can be found as part of the CFZ Blog Network, but also as a stand alone site at:

http://cfzwatcheroftheskies.blogspot.com/

Gulyabani

When we think of Islamic folklore, the first being that comes to mind is probably the jinni – a type of shape-shifting spirit that is mentioned in the Koran. But Muslims outside the Arabic countries have variant beliefs that are linked to their own pre-Islamic cultures. The people of Turkey also believe in jinn, but there is another figure with more ambiguous characteristics. This is the gulyabani, which is derived from the Arabic word "ghoul" and the Persian "yabani", which means wild or desolate. Put those together and you have a ghoul that inhabits desolate places such as deserts, ruins, and graveyards.

According to the Arabic conception, ghouls are the most vicious kind of jinn. The word "ghoul" derives from an Arabic root meaning

Joshua Allen

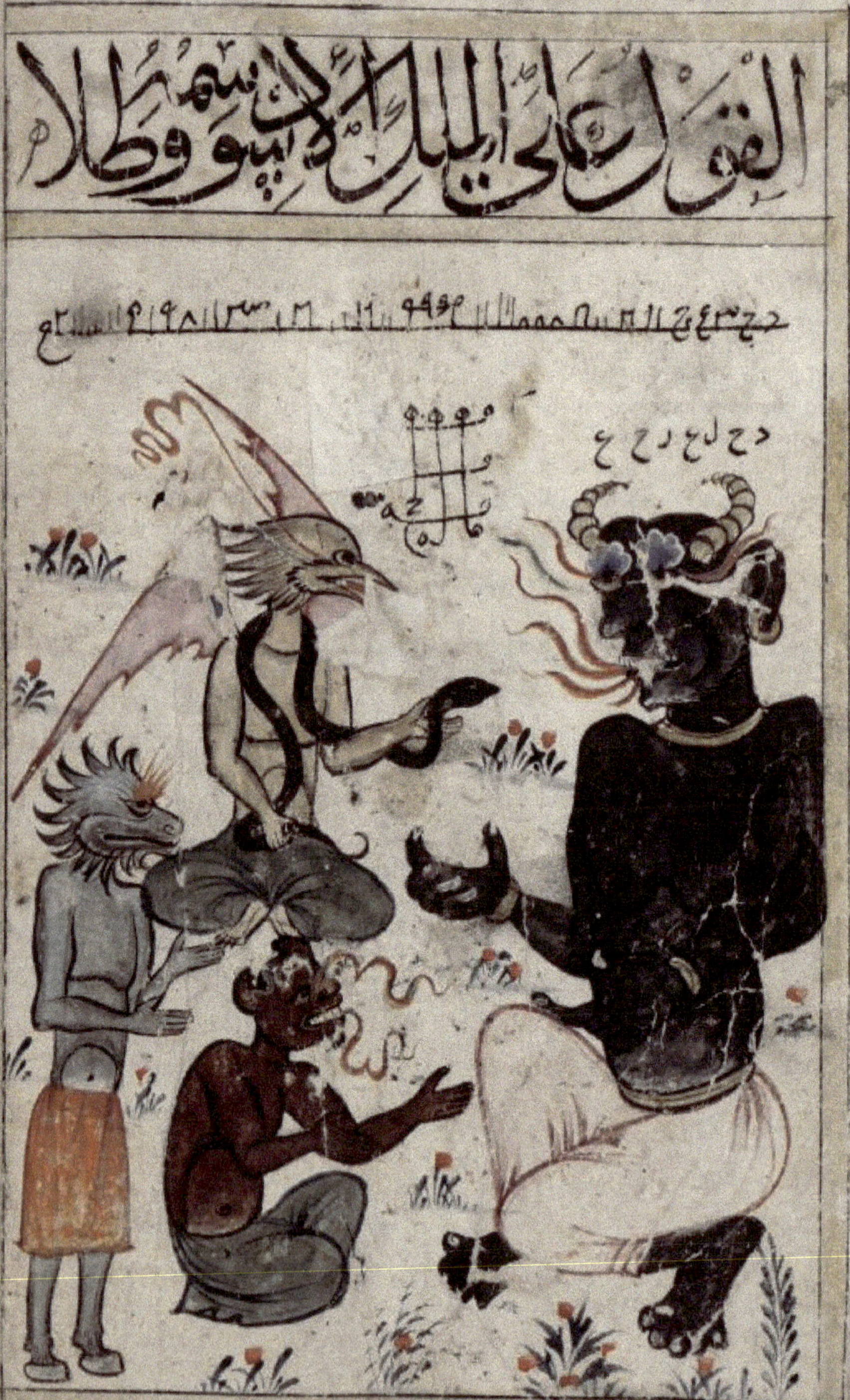
القول على الملك الاسود

"to destroy", which is what ghouls do to unsuspecting travellers. One of their tactics is to light fires on hills in the desert and lie in ambush, drawing people to their doom. In other accounts the ghouls appear as attractive women to lure men. But their physical characteristics vary widely, with one poet writing that a ghoul has the head of a cat with a forked tongue, the legs of a premature infant, and hairy skin like a dog's. A persistent feature of Arabic ghouls is their cloven feet, which they cannot hide even when they appear in a human body.

The idea of the ghoul entered Persian culture when the Arab Caliphate defeated the Sassanid Empire in the 7th century. Over time the Persian word "div" meaning "demon" came to be almost interchangeable with ghoul. A common feature in all Persian accounts is that ghouls like to eat human flesh, either living people or corpses that they dig up from graves. The most consistent physical description says the ghouls are tall, hairy creatures that emit an unpleasant smell. Persian poet Asadi Tusi wrote that the ghouls combine features of fairies and animals.

Despite popular misconceptions, the origins of the Turkish language, culture, and people are not primarily Arabic or Middle Eastern. The earliest written examples of a Turkic language are the 8th-century Orkhon stones in Mongolia. The common religion of all Turkic tribes was Tengriism, a shamanistic belief system whose chief deity was the sky. In the 10th century, a Turkic chief called Seljuq converted to Islam and migrated south into Persian territory. This was the start of the Seljuq Empire that went on to control Persia, Central Asia, Anatolia, and parts of the Levant. In this empire the Turks ruled over Arabs, Persians, Greeks, Armenians, Kurds, and other ethnicities. This was the cultural milieu that produced the Turkish version of gulyabani, which continued from into the Ottoman Empire and modern Turkey.

Descriptions of the Turkish gulyabani fall into two main categories. The first paints it as a hominid with reddish fur and backwards-facing feet. The other version is an extremely tall man with a long white beard. This second description was popularised by Ottoman writer Hüseyin Rahmi Gürpınar's novel "Gulyabani," published in 1913. This book later inspired the gulyabani that appears in the 1976 Turkish comedy film "Süt Kardeşler", and another with slightly better special effects in the 2014 film "Gulyabani". The bearded gulyabani has a ghost-like character that reflects Gürpınar's imitation of European gothic novels. With this slightly absurd gulyabani, Gürpinar was also aiming to satirise superstitions among the Ottoman public.

But there are interesting parallels between the furry version of the Turkish gulyabani and some other Asian creatures. Muslims in India, Pakistan, and Afghanistan report that the jinn have backwards-facing feet like the gulyabani. This characteristic is not found in Arabic accounts of jinn, which suggests that the reversed feet have a source within the Indo-Iranian peoples. Indeed, Muslims from these countries also use the word jinni to refer to "churel" – a female ghost with backwards feet that is part of Hindu folklore. In Nepal there is the Banhjakri, another furry creature with reversed feet. On the basis of this regional pattern, it seems likely that the Turkish gulyabani found its feet in the Persian world and then travelled with the Seljuqs into Anatolia.

There have been several reported sightings of a gulyabani in recent years. In 2007 the Hürriyet daily reported on a sighting in Fethiye on the Aegean coast. The witness in the village said that he and a group of friends heard cries and saw rocks being thrown from a hill. When they

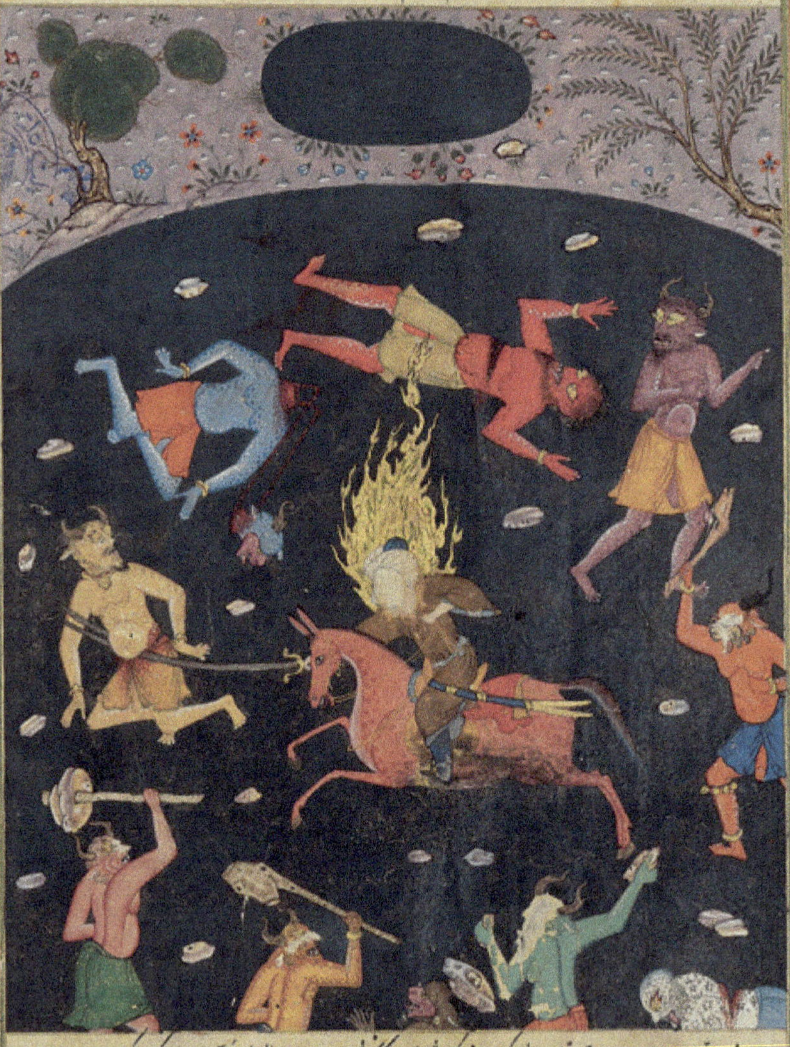

went to investigate, the witness saw a metre-tall hairy creature resembling a monkey coming towards him. Hürriyet claimed that villagers were keeping guard around the village into the night, until the gendarmes arrived and told them to go home. A sighting that conforms more with the novelistic gulyabani was reported in Sakarya province in 2013. Residents of Yeniköy village claimed to have seen a gulyabani that they called the "White Man" because of its long white hair and beard. This creature reportedly tried to force its way into houses and managed to escape twenty men in a cornfield. The local gendarmes were less concerned, stating that the nocturnal visitor might be a man wanting to see his girlfriend.

It's fair to say that a large number of Turks do not take creatures such as the gulyabani seriously. The novels, films, and news reports about the gulyabani tend to take a comical or sceptical approach. But with the lack of any concerted effort to investigate the creature, it is hard to say what truth the gulyabani might contain. This kind of research would also lead to a host of other Turkish creatures with roots in Greece, the Balkans, Anatolia, the Caucasus, and Central Asia.

Two regionally specific creatures are the demirkıynak of the Bigadiç mountains and the hırtık of the Fırat River. Others that are spoken of in broader areas of Turkey include the karakoncolos or winter jinni, the arçura or forest jinni, the azmıç or road jinni, the Laz or Georgian hominid known as Germakoçi, and a footless gulyabani called kayış. The roots of these creatures might be explained in the Ottoman archives, or they might be entirely within oral traditions. Even if the stories cannot be substantiated in any way, it is worth documenting them and releasing them into the global gene pool of potential creatures and oral tradition.

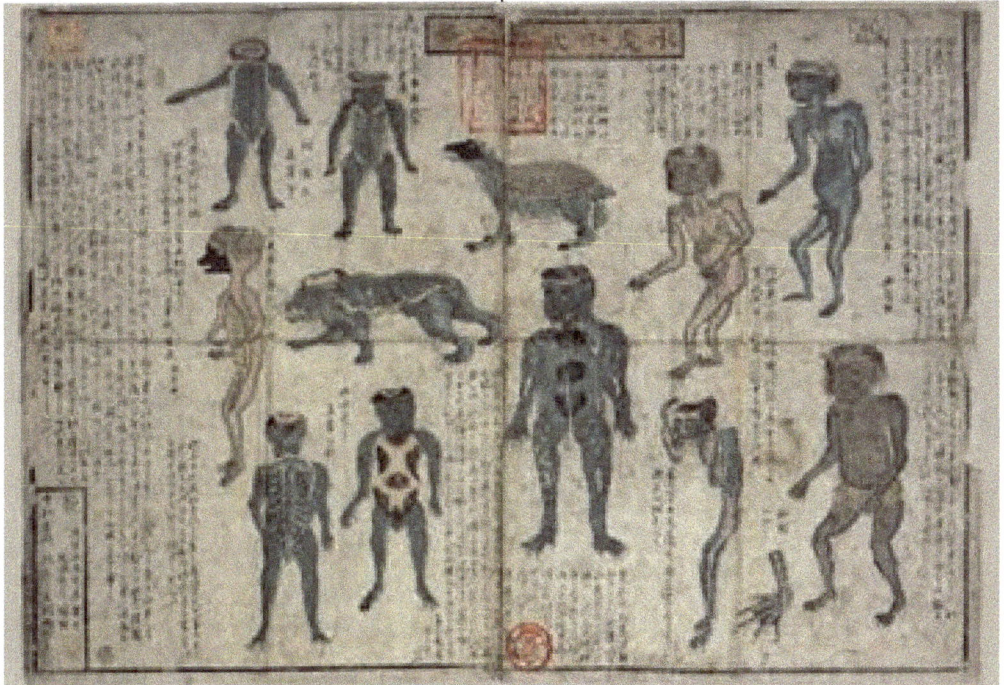

The Ohio Dogman

Ohio perhaps is best known for its unique museums and attractions, its wholesome inhabitants, and being the home state of a wealth of celebrities. Ohio is also well known in certain circles for the state's surfeit of unusual creatures and places. It is home to numerous ancient mounds, some of the most unique cryptids in the country, and a surplus of haunted residencies. But there is one creature which has been overlooked by most who examined the state's connections to the strange, Dogman.

Six to seven feet in height, it is described to appear like a cross between a person and a wolf or dog. Black, brown or grey fur covers the creature. Witnesses say that the creature appears similar to what a canine would look, had they the ability to walk for extended periods of time on their back legs. The creatures also seem to give off a presence of unnatural intimidation. The following are four examples of the Dogman's activity in Ohio.

The Defiance Werewolf

Defiance, Ohio, a sleepy little town, about 55 miles south of Toledo, had a series of unusual encounters in 1972 that will become whispered legend in the years after. Defiance is your average, small, Midwestern American town. Most of the people who live there make their living by working at one of the handful of factories around the area. Defiance is best known for its Fort Defiance, which has a colorful historical significance.

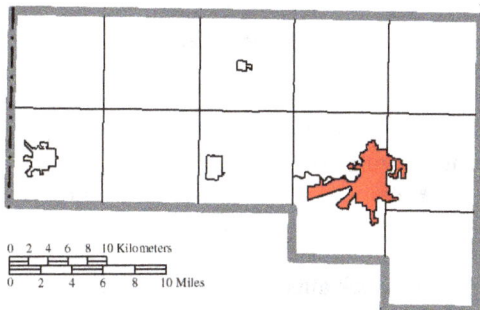

0 2 4 6 8 10 Kilometers
0 2 4 6 8 10 Miles

So what happened in Defiance all those years ago? Possibly the oddest couple of Dogman encounters ever reported.

On July 25th, 1972, two employees of a local freight train company were working the graveyard shift, inspecting railcar braking systems. Ted Davis, one of the witnesses, during an interview with the *Toledo Blade* said, "I saw these two hairy feet. Then I looked up and he was standing there with a big stick over his shoulder. When I started to say something, he took off for the woods." The creature was also described as having, "huge, hairy feet, fangs, and it ran side-to-side like a caveman in the movies."

About a week after the first encounter, Davis claims to have seen the creature, at the same place as the first encounter, the rail yard. He said it looked at him from the edge of the woods, and then turned and left. That same night, a grocer reported an attack from a hairy, animal-headed creature wielding a two-by-four.

In a small town, nothing stays quiet for long.

About a week after the initial events, the local newspaper caught wind of the unusual story. The headlines wrote "Horror Movie Now Playing on Fifth St." and opened with the famous poem from *The Wolf Man*:

"Even a man who is pure of heart,
And says his prayers by night,
May become a wolf when the wolf bane blooms,
And the moon is clear and bright."

The Crescent News, the local newspaper, and the *Toledo Blade*, ran a total of four articles about the incident. The longest of the articles focused on the local police force's investigation into the incident. Headed by Donald Breckler, the police chief at the time, the police searched extensively for the "werewolf". Although the general public thought it to be a bored teenager's prank, Chief Breckler considered the club-wielding, animal-faced assailant was a threat to the community. Eventually, after days of looking and no results, the search soon ended.

It is worth noting that Defiance, Ohio is no stranger to unusual phenomena. Within the past fifty years Defiance has seen hundreds of Unidentified Flying Objects, a Dracula lookalike running amuck, and numerous other tales of the weird.

Akron's Dogman

Summit County Metro Parks cover about 22 miles of land within Summit County, Ohio. This space is largely used for hiking and biking trails, local parks, and the preservation of Ohio's diverse flora and fauna. A place such as one of these trails and parks would be an opportune place to observe wildlife, and maybe something else?

Since 2011, there have been reports of people encountering six-foot tall, bipedal, wolf-like creatures running amuck in the Metro Parks. The rash of sightings began in January, 2011 when multiple people reported seeing a creature dash across the road in front of their cars. They described the creature to as a human-sized dog which ran upright, on its hind legs, like a person.

Later that same month, a park ranger had a sighting of a similar creature. According to his story, as he was making his daily rounds he saw a figure which appeared to be a dog laying in the snow. He got into his vehicle and shined the headlights on the figure which then rose, looked at him, and walked away. The ranger returned the next morning with researcher Joedy Cook. There they found 168 tracks which were human in shape and size, but had sharp indents at the front of the foot.

Silver Creek Dogman

In the fall of 2013, a witness who went by the name Andrew, contacted researcher Ken Summers about two encounters with unusual creatures in Norton, Ohio.

Andrew was returning home after work, across Johnson Road, which is on the border of Silver Creek Metro Park in Norton, Ohio.

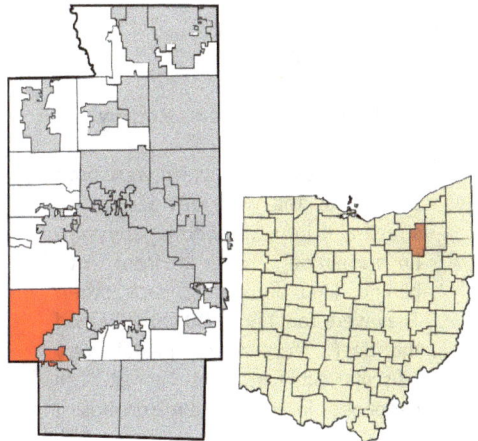

He was nearing an intersection when he stopped his car because two deer were racing across the road. He saw creatures that he could not identify following the deer. He also said that what he saw did not resemble a Bigfoot.

Here is his description;

> "I would place them somewhere between 6'6" and 7' tall. They chased the two deer (which were both smaller, by the way) out across the street and into the woods. They ran [in formation]... one in front, two behind, kind of next to each other. They were roughly 30-40 yards behind the deer. They were bipedal, very muscular, and fast. Lightning fast. It all happened in just a few seconds. I

> couldn't describe any features, unfortunately; I'm assuming it was either a new moon or cloudy because it was very dark, but they were definitely a dark color... maybe a chocolate-brown or a black color."

A few weeks after the initial encounter, again while travelling along Johnson Road, Andrew saw something run out of a cornfield in front of his car. His account follows;

> "The fields had full-grown cornstalks, but I don't know exact heights only that the cornstalks were taller than me by a head and I'm 6 feet tall. This time, [my sighting of two creatures] was a quick flash because there was

no open land to it. They basically leapt the road as they broke the corn and landed about 10-15 feet into the field on the other side and kept running. This time, the pair that I saw in the moonlight, the first was black and the second was black with white or silver on its chest and back.... Since the first three I saw were all a solid color that means there must be at least four different creatures."

A Camper's Story

A very interesting encounter occurred a few years ago in Adams County, Ohio, during which a group of campers encountered a dog-like humanoid. I was asked to keep the witnesses' names confidential.

The following are excerpts from an email I received where one of the witnesses explains the encounter;

"When we got there, we, the kids, were just exploring for the first day and we were learning our terrain. We walked around our area which was a good 10 square miles, while we were

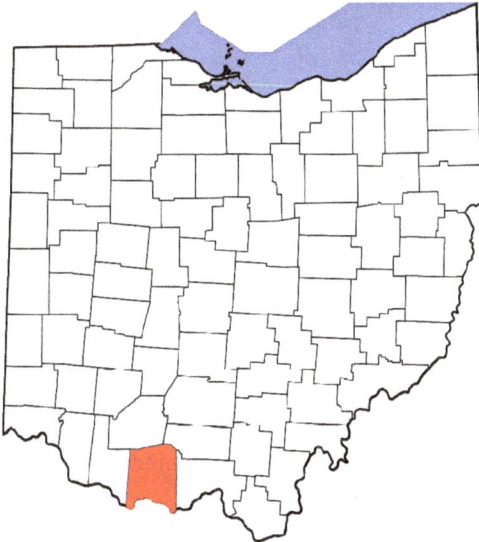

out there while it was nearing dark I started to get uncomfortable and notice things around us that seemed out of place. I also felt as if someone or something were watching us, so we went back to camp.

The campground was a semi-circle in shape with a semi steep hill going down on the curved side of the semicircle. On the flat side was a road that separated us from the forest but it wasn't very large only about a car width in size. The semi-circle was pretty large, we had a few tents set up and we still had plenty of room. The fire pit was pretty much dead center of the camp. If you were facing the road at the fire pit you couldn't see anything for about 30 degrees because of the trees.

While cooking dinner, we heard branches snapping across the road. We assumed it was just an animal wandering around, and they eventually stopped.

My father stood up and got out a searchlight and turned it on and started to walk around the perimeter with my friend and I. Once we got about 3/4 down the road I noticed a pair of eyes looking at us from the trees. My father didn't notice them and continued sweeping but I stopped him and made him point his light back towards the eyes. My father shined the light at the creature, which was about seven feet tall, was covered in black fur, and looked like a cross between a wolf and a human. After about ten seconds, the creature ran into the woods.

While everyone else slept in the car,

my father and I remained in our tent. My father then said that at around midnight he heard the creature come into camp and sniff around the car and our tent. After a little bit, the creature disappear into the other side of the woods and heard the creature disappear. He didn't go to sleep the rest of the night.

The next morning, as we packed up camp, my dad the saw a black dot on one of the large hills about a quarter mile away. We got out the binoculars and saw that the black dot was the creatures from the night before, laying like a dog. It looked like it was watching us. Needless to say, we got out of there very fast."

This encounter is interesting because this is, to my knowledge, the first story in which a Dogman stays in the area of the sighting, and watches the witnesses. It is almost as if it was waiting for them to leave its territory.

Conclusion

The encounters above are just four examples of the widespread Dogman phenomena in Ohio. The Native Americans in the area had a strong connection with the wolf. A series of mounds in Southern Ohio are even named after the animal. On the Internet, numerous people report having encounters with bipedal canines.

What have people been seeing in Ohio? Misidentifications of known species seem unlikely. The last of Ohio's wolves was killed in 1824, and, in any case, canines don't normally walk on their hind legs. Some researchers have associated

sightings with the "zooform" (a creature which appears to be an animal but are supernatural in origin) phenomena. In truth, no one knows what these creatures are. We simply do not yet know enough. The mystery of Ohio's Dogman is truly a strange one, proof that we don't yet know or understand every shadow that stalks our parks, forests, and roads.

References

Newspapers
- Armstrong, Ellen. (1972, August 2). Horror Movie Now Playing On Fifth St. *The Crescent-News*.
- James Stegall. (1972, August 2). Werewolf Case in Defiance Not Viewed Lightly By Police. *The Toledo Blade*.

Websites
- Ohio's Silver Creek Dogmen: Investigating a Frightening Encounter With Midwestern Werewolves. (2014). Retrieved April 26, 2016, from http://weekinweird.com/2014/12/08/exclusive-ohio-silver-creek-werewolf-investigation/
- Werewolves in Ohio? - Lunar Falls. (n.d.). Retrieved April 26, 2016, from http://lunarfallstrilogy.weebly.com/werewolves-in-ohio.html

Books
- *Dogman: Werewolf Encounters in Northeastern America* by Joedy Cook
- *Real Wolfmen: True Encounters in Modern America* by Linda S. Godfrey
- *On Dogman* by Ronald L. Murphy Jr.

THE YETI:
AN APE NOT A BEAR

Recently much has been made in the media of supposed yeti hair being identified as coming from an 'ancient strain of polar bear' believed extinct for 40,000 years. Many thought that the mystery of the yeti had been explained and that the legendary 'man-beast of the Asian mountains' was nothing more than a strange kind of bear. But, as anyone who has given the mystery even the most cursory of examinations will realise, this simply does not fit the accounts and the yeti has always been seen as some form of higher primate, an ape or 'wildman'.

In this article we will take a closer look at this case and try to find out what actually happened.

Professor Bryan Sykes is one of the world's leading geneticists, based at Oxford University. In 2012 he set up the Oxford / Lausanne Collateral Hominid Project with Dr Michel Satori of the Musee de Zoologie in Paris. The project collected supposed hair from the yeti and other mystery primates. The project was filmed for Channel 4 in a three part series entitled *The Bigfoot Files*.

Prof Sykes conducted the DNA tests on hairs from three unidentified animals, one from Ladakh - in northern India on the west of the Himalayas, one from a supposed stuffed yeti shot in Tibet, and the other from Bhutan. The results were then compared with the genomes of other animals that are stored on a database of all published DNA sequences.

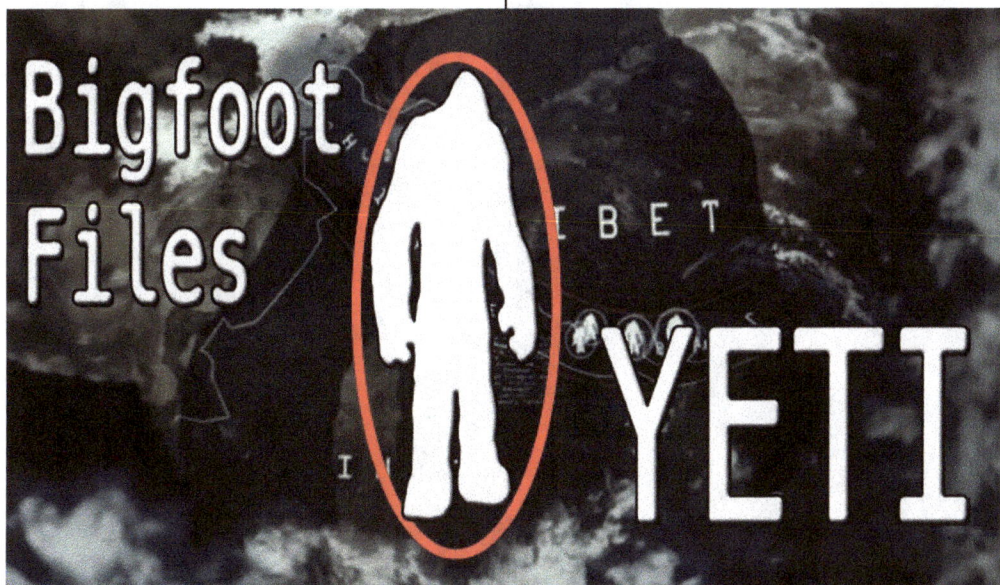

RICHARD FREEMAN

The Ladakh specimen was collected from a mummified body. French explorer Christophe Hagenmuller was shown the body of a creature shot some thirty years before and kept in a remote village. He described the body as having both bear-like and wolf-like features and being the size of a small man. The hunter who killed it said he did not think it was a bear, but it had no primate features either.

The Bhutan sample was collected by a film maker some ten years ago. It was found in a hollow tree whilst the team were making a documentary about the yeti for the Channel 4 series *To The Ends of The Earth*. The professor had tested the DNA from this specimen before but could not match it to anything on his database.

The stuffed 'yeti' was held at Sigmunddskron Castle in Bolzano, South Tyrol in the Italian Alps. The castle is now a museum run by Italian mountaineer Reinhold Messner. It was shot in Tibet in 1938 by Ernst Shafer who was leading a Nazi expedition into the area to investigate the idea that Tibet was the ancestral home of the Aryan race. The creature was shipped back to Germany. On Shafer's death his widow donated it to Messner. The creature was transparently a very badly stuffed bear and no kind of primate whatsoever.

The Nazi taxidermy yielded no DNA at all, but both the Ladakh and Bhutan samples did and what's more they matched each other. They were also a 100% match for something surprising, DNA from an ancient strain of polar bear at least 40,000 years old. The DNA was originally said to match that of a jawbone from a prehistoric polar bear found on the Norwegian island of Svalbard. It was genetically quite distinct from the Himalayan brown bear *Ursus arctos isabellinus*.

The polar bear *Ursus maritimus* is not native to the Himalayas. However the polar bear speciated from the brown bear *Ursus arctos* about 603,000 years ago; geologically speaking, quite recently. One theory was that the Himalayan mystery bears represented a surviving form of this early strain. Another is that they were polar bear/brown bear hybrids. The two species are known to interbreed on occasion in places where their ranges overlap. The Himalayas are not one of these regions though.

Later investigations by researchers from the University of Copenhagen have refuted Sykes' claims.

The samples, just three, only two of which yielded DNA, are very small. However you might think that this wrapped up the mystery of the yeti. You would be quite wrong.

Of the three samples taken two were known to have come from bears or bear-like creatures, not primates.

The Bigfoot Files only interviewed one local man. He had not seen the yeti but had heard strange calls and found two of his yaks devoured. When Irish explorer Peter Byrne was conducting expeditions for the Texas oil millionaire Tom Slick in the 1950s he interviewed many eyewitnesses. He showed them pictures of apes such as gorillas and orang-utans, pictures of bears and artistic reconstructions of hominans such as *Homo erectus* and Neanderthal man. The witnesses invariably chose the gorilla as being closest to what they saw. They said, however, that the creature walked on two legs not four. It seems as if the producers of *The Bigfoot Files* do not want eyewitnesses describing something that contradicted their bear theory.

Reinhold Messner himself is an interesting character. In the show he is interviewed and claims that he saw a yeti in a Himalayan forest at night. His description is vague 'a large, dark being'. He later found bear prints that he said looked like the famous prints photographed by Eric Shipton in 1951 on one of the glaciers of the Menlung Basin, north Tibet. In fact they in no way resemble the Shipton picture, least of all in the fact that the bear prints show obvious claw marks which are lacking in the Shipton shot. Even when a bear's print is doubled up by the animal treading in its own forepaw print with its hind claw, the result looks precious little like any supposed yeti track. He said that the locals referred to it as 'chemo' the native name for a large, rare kind of bear.

Messner authored a book *My Quest for the*

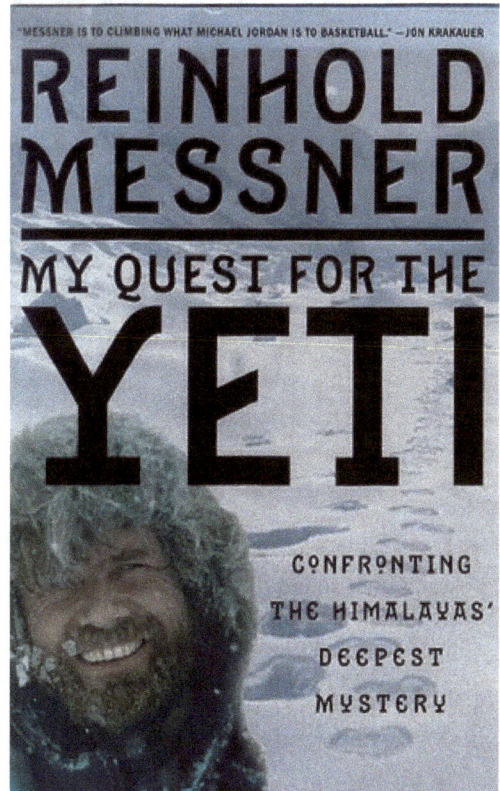

"MESSNER IS TO CLIMBING WHAT MICHAEL JORDAN IS TO BASKETBALL." — JON KRAKAUER

REINHOLD MESSNER

MY QUEST FOR THE YETI

CONFRONTING THE HIMALAYAS' DEEPEST MYSTERY

Yeti in 2000 in which he pushes the bear identity. This is curious as previously he had described the creature as a primate-type animal.

I once interviewed the actor Brian Blessed, a renowned explorer and mountaineer himself, for a long defunct and not very good magazine called *Quest*. Blessed, who is a friend of Messner, said that he had told him of his encounter with a yeti. Blessed said that Messner had walked around some rocks and come 'face to face' with the creature. He said it was not a bear, was 7 feet tall, man-like and stood erect.

There are other occasions when Messner's descriptions sound precious little like a bear. Julian Champkin of the *Daily Mail* 16th August 1997 wrote that Messner has... "encountered the yeti; and not once, but four times, once close enough to touch it. More importantly, he claims to have photographs of the creature, including a mother yeti tending her child, and a yeti skeleton".

Needless to say none of his pictures have been forthcoming. Messner goes on... "I searched for a week, 12 hours a day, in an area with no trees," he says. "I didn't expect to find one so soon. First, we saw a mother with her child. I could only take a photograph from the back. The child had bright red fur, the older animal's fur was black. She was over two metres tall, with dark hair, just like the legend. When they saw us they disappeared."

Two days later, he claimed to have come across and filmed a sleeping yeti. The film is just as noticeable as the photos by its absence.

In an article relating to the BBC's *Natural World* documentary on the yeti, Messner describes seeing one from a range of 30 metres in Southern Tibet. The article says Messner is sure it is some kind of primate. He describes it in the article thus... "It was bigger than me, quite hairy and strong, dark brown-black hair falling over his eyes. He stood on two legs and immediately I thought he corresponds to the descriptions I heard from Sherpas and Tibetans."

So why did Messner write a book trying to explain away the yeti as a bear when this transparently was not the creature he claimed to have seen? Was it because of fear of ridicule? And what became of the photos and film? Was Messner trying to take the focus away from these or make them seem less important by saying the yeti was just a bear? Could this be because the film and photos did not exist?

I'm inclined to dismiss Messner's claims of the yeti being a bear in the light of these past statements and his seemingly strange change of tack.

In 2010, I went in search of the yeti in the Garo Hills of Meghalaya in North India. Here the yeti is known as Mande barung or 'forest man'. I interviewed many witnesses from the hill tribes, some of whom had seen the creature as recently as a year ago. All described the same animal. Ten feet tall, covered in black hair and walking erect. The said it looked like either an upright walking gorilla or a colossal, hair covered man. All were adamant that the creature was not a bear, an animal they were very familiar with. One man saw the creature building a nest like apes do. Another saw a female eating bamboo whilst suckling a youngster. Sketches they produced showed ape-like creatures that in no way resembled bears.

I also came upon some man-like track in the earth next to a stream. These displayed five toes and were 12 inches by 5. They were driven into the mud to a depth of 2.5 inches. The creature seemed to be following the stream and overturning rocks to look for fresh water crabs. Local researcher Dipu Marak told us he had seen tracks 19 inches long in the Garos.

The mountain forests of the Garos stretch off into Assam where the yeti is known as konglangpo and continue into the wilderness of Bhutan where the beast is known as migo. A detailed account of the expedition may be found in the book *CFZ Expedition Report: India*.

Further confusion may come from the Tibetan name 'dzu-teh' meaning 'hulking thing', a name used to describe both bears

and the true yeti. Another myth that needs exploding is that the yeti is white. There has never, ever been reports of a white yeti. The creature is said to be black, brown or ginger in colour but never white. The sorry idea seems to have stemmed from a mistranslation of the Sino-Tibetan name 'metoh-kangmi' meaning abominable man of the rocks. It was mistranslated as 'abominable man of the snows' in Bill Tilman's 1938 book *Mount Everest*. The creature lives in the forests well below the barren snowline. The word 'yeti' is Nepalese and means 'rock-beast'.

CONCLUSIONS
Reports of the yeti show characteristics that cannot be attributed to bears. These include the following.

- ### STANCE
All bears, living or extinct are quadrupeds, they walk on all fours. Bears are capable of standing and walking erect for short periods but seldom move for more than a few steps on two legs. An erect posture in bears is usually a threat, making themselves look larger to intimidate enemies or rivals. Bears may also stand erect to reach food items.

Rarely, individual bears will take to moving on two legs for longer periods due to injury. However they move with mincing steps and an awkward gait quite unlike the naturalistic, man-like walk of the yeti.

- ### SHOULDERS
Like all quadrupedal mammals the scapulae or shoulder blades of bears lie flat against the sides of their body. The scapula of humans jut outwards giving us our characteristic broad shoulders look.

Apes, too, despite mostly being knuckle walkers have jutting scapulae and broad shoulders. A bear on two legs lacks the distinct broad shouldered appearance. One of the most notable things about descriptions of the yeti are the remarkably broad and massive shoulders.

• OPPOSABLE THUMBS

Bears have five non-retractile claws on their paws. They have no opposable or semi-opposable thumb. The giant panda, *Ailuropoda melanoleuca* has a pseudo thumb in the form of an enlarged sesamoid bone which it uses to brace and manipulate bamboo stalks. This feature is not found in

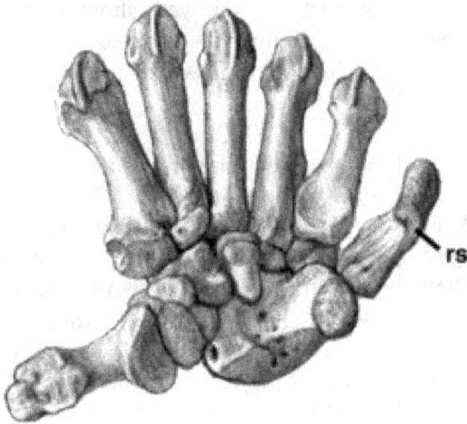

bear, *Arctodus simus* and *Arcodus pristinus* inhabited North America and Mexico until around 11,600 years ago. However despite having snouts shorter than any modern bear they in no way resembled apes. Short faced bears were confined to the Americas and shared all of the other ursine features that make bears poor candidates for the yeti.

• METHOD OF KILLING YAKS

Bears kill yaks and other large animals by biting and clawing. The yeti is said to kill yaks in several ways. Firstly by grasping the horns (with gripping hands that bears do not have) and twisting the head till the neck brakes. Secondly by punching the forehead

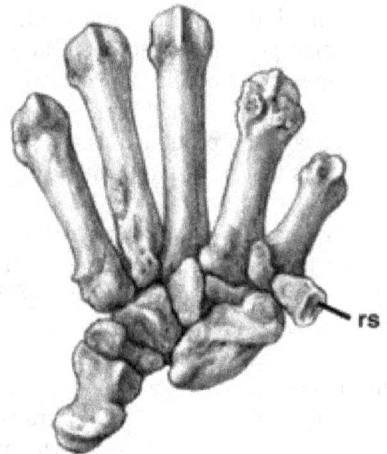

any other species of bear.

The yeti is said to have man-like, grasping hands and can manipulate objects. The yeti is said to pick up and hurl sizeable rocks, something no bear, not even the panda, can do.

• FLAT FACE

The yeti has a brow ridge and prognathous jaws. Its face however is flat when compared to the faces of bears with their dog-like snouts. Two species of giant short faced

with such force that the skull shatters and pierces the brain and thirdly by smashing the head with a rock. Again this is clearly not the behaviour of a bear.

All of the above point to a primate, probably a great ape unknown to science or of a species believed to be extinct, but known from the fossil record. The discovery of an ancient form of bear in the Himalayas would be fascinating and exciting to cryptozoologists in and of itself, but it can in no way be used to 'explain away' the yeti.

~~ZOOLOGICAL ODDITIES~~
IN FOREST AND
STREAM MAGAZINE
VOL 1.

PART ONE

1873 – 1874

LODGES OF THE BLACKFEET === MATANZAS WAY
IN THE TREE TOP WITH A CAMERA

VOL. LXVI.—No. 25.　　　　PRICE, TEN CENTS　　　　SATURDAY, JUNE 23, 1906.

Copyright, 1906, by Forest and Stream Publishing Co.　　FOREST AND STREAM PUBLISHING CO., 346 Broadway, New York　　Entered at the New York Post Office as Second Class Matter

Forest and Stream was a magazine featuring hunting, fishing, and other outdoor activities in the United States. The journal was founded in August 1873 by Charles Hallock. At the time of its 1930 cancellation it was the ninth oldest magazine still being

RICHARD MUIRHEAD

issued in the US. [1]

Here I present a fairly comprehensive summary of some interesting Fortean Zoological articles from *Forest and Stream* which follows the tradition of Hardwicke`s Science Gossip; which I had published in the *Centre for Fortean Zoology Yearbook* in 2008. I cannot vouch for the accuracy of everything here, nevertheless I hope you find it interesting. I used Biodiversity Heritage Library`s online archive to locate *Forest and Stream.*

VOLUME 1
1873

August 28th 1873 p. 38 **HEDGEHOGS SUCKING FROM THE TEATS OF COWS.**

" Is it possible that the same story exists in the United States as in England as to Hedgehogs sucking cows? Mr Frank Buckland wants more evidence about it. Here is a correspondence on the subject:

" A dispute has arisen consequent on my defending the hedgehog from the charge of sucking from the teats of cows, an idea, I thought, long since exploded. It was finally arranged that the matter should be considered as finally settled by your decision. I may say that one individual, who backed his opinion with a wager, swears most positively that he has seen the cow rise from the ground and walk away two yards with the hedgehogs still clinging to the teat – A.B. [It is a curious thing how these old fancies crop up from time to time. If a hedgehogs mouth is examined, it will be seen that it is much too small to take in a cow`s teat. Hedgehogs are very fond of milk and I think it is very likely that they will lick

up from the teat any milk that is exuding – hence the origin of hedgehogs sucking cows, I should be glad to receive more evidence on this point." FRANK BUCKLAND.

European Hedgehog foraging at night. Wikipedia Creative Commons Share-Alike 3.0 Unported. Samuel Palmer originally took this photo.

August 28th 1873   p.38   **LONGEVITY OF ANIMALS**

"Highlanders believe that the deer is the longest lived of all the creatures, save the eagle. They have an old Gaelic proverb which is worth recording:

Thrice the age of a dog is that of a horse;
Thrice the age of a horse is that of a man;
Thrice the age of a man is that of an eagle;
Thrice the age of an eagle is that of an owl."

The exact longevity of animals has never been properly determined, and is a subject worthy of special attention.

We have seen an English horse in Canada thirty-three years old. We would like to have some authentic data on these subjects."

September 25th 1873   p. 99
VENOMOUS LIZARDS IN CANADA
HALIFAX BARRACKS. EDITOR OF
FOREST AND STREAM.

"Having noticed your article on "Woorai" where Dr Saffrey describes a venom as coming from a frog used by the Indians to poison their arrows, I beg leave to state the following facts, which I trust will find some explanation in your columns: Two years ago whilst hunting in the mountains of Nova Scotia with Indians, during the middle of September, in the dense hard wood forest, I came across a rather large lizard. I was about securing him, when my Indian cried out " retez!" "retez!" their French for stop. On inquiring the reasons for their caution, they said it was a most dangerous creature calling it "the man poisoner lizard."

They assured me that if there was a scratch on my finger, and I touched the lizard, it would poison me fatally. I cut a stick and poked him. I did this easily, as his movements were rather lethargic. This seemed to enrage him, and he turned on the stick. At the same time a peculiar white creamy and glutinous matter exuded from his body. One of the Indians caught a field mouse . I touched the mouse, who was quite lively, and unhurt, somewhere about the mouth, with the stick which was imbued with the exudation of the lizard, and the mouse died in violent convulsions in a few minutes. I regret very much not having preserved the lizard, which I killed...I found out afterwards, that the settlers all believed that this lizard was terribly venomous and cited accidents arising from touching it. The colour of the lizard was greenish yellow. His body was about four inches long ,and his tail about five inches. He had two quite sharp teeth, fangs in fact on both his upper and lower jaw, and smaller teeth in front. Have any of your readers come across a similar lizard? Is there any foundation in the universal dread people have of toads and lizards? I think there may be some reasons for it."   CAPTAIN P.

[There are some 5000 different species of lizards worldwide and, until a few years ago, only two (the Gila Monster and the Mexican Beaded Lizard) were thought to be venomous. Scientists also believed that those two species had evolved venom production independently from snakes.

Indeed, until recently, nasty swelling and excessive bleeding resulting from another lizard's bite was thought to be due to infection from the bacteria in the reptile's mouth.

But a team of researchers from the University of Melbourne, Australia have revolutionized herpetology by showing that venomous lizards are actually much more widespread than thought. These scientists, under the leadership of Bryan Fry, have demonstrated that both monitor lizards (commonly kept as pets) and iguanas also produce venom. Nine types of lizard toxins are shared with snakes, but some toxins are new and yet to be investigated for medical research.

Furthermore, it is now thought that venom production had, actually, a single early origin for lizards and snake and that the common ancestor to all venomous species lived about 200 million years ago. The evolution of venom would have, thus, coincided with the rapid spread of small mammals.

To date, the toxin-producing oral glands have been identified in species of the anguimorph and iguanian lineages. It is believed that as many as 100 species of living lizards actually use venom.]

VOLUME 2
1874

March 5th 1874 p. 57   **MEN WITH TAILS** (A classic piece of Forteana here, covered by the late William Corliss, amongst others.)

EDITOR *FOREST AND STREAM*
"At the risk of injuring my reputation for

veracity, I propose to furnish the Forest and Stream with an account which, however strange, is actually true, of some people I have met with.

Fortunately I have among my friends, and you have among your readers, some of the most learned men in the land – who are at the same time – possibly because they are so learned firm believers in ,and supporters of the Darwinian theory of evolution and development, and I trust to them to come to my rescue with argument if my facts seem a little to strong. Briefly, I have in my knocking about the world, *met two persons with tails.*

Originally appeared in October 1952 issue of "Astonishing" # 18

One of said tails I saw distinctly, the other, as distinctly…The one I saw was in Africa. A party of four of us started early one bright forenoon to drive from Widdow`s Hotel in Cape Town , to a sunny little English town at Wynberg where we proposed to enjoy an out of door dinner and return in the evening. About three or four miles from Cape Town we passed – as we had several others - a little collection of Hottentot huts, located under a coconut grove on the left side of the road. A number of little children, all naked, were playing between the road and the huts, and as we approached and passed, scuttled off rapidly for the huts. One little fellow – boy or girl I don`t know which – about six or eight years of age, was not twenty feet from us ,running, and we saw plainly that it had a prolongation of the spinal column about two or three inches in length. This prolongation was pointed, something in size or shape like a very taper finger and pointed nearly straight down...

In the other case, I was the guest of Mr John Mitchell, a prominent resident of Pulo Penang , an island in the Straits of Malacca. A servant of the family was a woman of perhaps forty years of age, called Mary Andaman. She was a native of the Andaman Islands ,and had been brought to Penang when but a little child. At that time Mary had a penchant for doing without clothing and it was a well known fact that she had so much of a prolongation of the spinal column that it was popularly said that she had a tail.

I heard this from several and asked Mr Mitchell about it. He said that it was true, but said it was a subject seldom spoken of; that the woman, who was a respectable woman, and a member of the church, was very sensitive in regard to the manner, and for years had refused to submit to any

examination, and was annoyed and angered by reference to it. I saw the woman and have perfect confidence in Mr Mitchell."
PISECO.

March 19th 1874 p. 86   **STRANGE ANIMALS OF THE WEST**

RUSSELL,KANSAS, February 18th, 1874

EDITOR *FOREST AND STREAM*

"I wrote to you some time ago asking the name of a rat I had found here. At the same time I wrote to a friend in Junction City about it. As he and you do not agree and he also mentions several others I will make some extracts from his letters and you can publish them if you wish. Mr Green or "Senaca Bill" as he is known here ,has been trapper and Indian trader on the plains and mountains for over thirty years…He says "the rat you speak of I call the Carrier Rat (from his habit of carrying all sorts of old trash to his nest). The head of the Republican is as far north as I ever saw him, and never east of the Missouri. It is also found in Texas and New Mexico and in Mexico very plenty; but I never seen it in Colorado. In the mountains there are four more kinds. One is somewhat larger than the Carrier Rat; black on the upper parts, underneath yellow; no ears visible, and a very short tail. Another is blue all over except nose, feet and tail which are black. The tail of this kind is very long; its bite is poisonous. Another is white, and is as large as a muskrat. Then the great Kangaroo Rat also inhabits the mountains…" I know of five different kinds of squirrels in the mountains and none of them as large as the Eastern chipmunk… There is also a very large kind with wattles under the throat."

"I had been told that the polecats of this country were poisonous, and asked him about it. He said " I knew a man to be bitten on the thumb by one and he died on the third day after. This was on the Jacques River. In Dakotah, about fifteen years ago. I have also heard of three persons poisoned to death in like manner in Kansas. Nothing but an all fired big drunk can save a man, or something to counteract the poison immediately and thoroughly."

April 2nd 1874 p. 122 **GIGANTIC OCTOPUS IN JAPAN**

"Mr W.B Tegetmeier of the *Field* , gives a most clever account of the gigantic octopus in Japanese waters, taken from a work called " Land and Sea Products" by Ki-Kone, who must be a naturalist of the most distinguished merit. The text is illustrated by two amusing wood cuts, which have that peculiar matter of fact treatment which render Japanese pictures so quaint and yet truthful. Before Mr Harvey`s most accurate description of the octopus produced by the FOREST AND STREAM, with a picture of the tentacles of the *monstrum horrendrum* , its existence was almost doubted. One of the Japanese pictures shows a boatman fighting with an octopus and lopping off his arms with a big knife…In the second picture these material Japs have the octopus cut up into bits for sale at a fish stand, with groups of admiring spectators ,pretty much the same as one sees at Mr Blackford`s ,in Fulton market ,when he shows his brook trout off in his aquarium to an admiring crowd…"

April 23rd 1874 p. 163 **FROGS CRYING LIKE A CHILD**

ARLINGTON,MASS., April 10, 1874

EDITOR *FOREST AND STREAM*

"I notice in Forest and Stream of April 2nd an item stating that frogs, if subject to torture [don`t try this! – Richard], will shriek like a child. The above statement I can fully corroborate. While walking near a pond last

summer I surprised a large bull frog (*Rana pipiens*) (actually the Northern leopard frog) some distance from the water. On cutting off its retreat to the pond and whilst poking it with a stick, it uttered sounds resembling the cries of a frightened child. Continuing my annoyance its cries of fear changed to shrieks of rage, and it bit repeatedly at the stick. It would be interesting to know whether this is the usual behaviour of the creature under like circumstances."   MERLIN

May 21st 1874. p. 3 missing; p.
234 **THE DODO THAT WASN`T A DODO**

THE DODO – "Dr A.B. Steinberger whose recent visit to The Navigator Islands [Samoan Islands] is the subject of a highly interesting report to the Secretary of State ,made a discovery in the course of his explorations upon which he prides himself greatly, relating to the present existence of the dodo, a bird long supposed to be extinct .It is the tooth billed pigeon, having three teeth upon either side of the lower

mandible. The doctor has brought home a living specimen and also a dead specimen preserved in spirits. In regard to its habits he says that " it is a timid bird , lonely in its habits , exceedingly scarce in number, and only found and almost inaccessible parts of the mountains:" and in regard to the value of his discovery he says:

"For about two centuries past the few remains of this bird known to the scientific world, as a foot or a head , together with some paintings made of it in the seventeenth century, have been preserved in European museums with great care, and have been regarded as of great value. Several scientific treatises upon it have been contributed to learned societies within the last fifty years , but so little has been known of it by naturalists that they have not been agreed as to what family it belonged , or what were its habits. Ornithologists will doubtless recognise in these specimens an extremely valuable acquisition to science, and improve the opportunity now offered of making a thorough acquaintance with its peculiarities and of determining its proper classification."

May 28th 1874. Page 244. **THE DODO** SMITHSONIAN INSTITUTE , WASHINGTON D.C

EDITOR *FOREST AND STREAM*

"On page 234 of your issue of May 21st I notice an article upon the dodo, stating that this bird recently had been captured alive in the Navigator Islands by Dr A.B. Steinberger. The bird which this gentleman secured was not the dodo, but the dodo pigeon (*Didunculus strigirostris*) a species which is well known to ornithologists and which exists in more than one museum, the National Collection having had a good mounted example for many years. Though a member of the pigeon family, the present species is so aberrant that Bonaparte instituted in 1850 a separate family (*Dinduculidae*) for its especial accommodation ; but its characters scarcely warrant so wide a removal from the true pigeons and Gray`s sub-family (*Didunculniae*) may be accepted as a probably more truthful expression of its relationship.

The generic name of the dodo pigeon (*Didunculus,*) signifies "little dodo" and was suggested by the resemblance of many parts of the bird to those of the dodo, (*Didus ineptus.*) It is perhaps needless to add that the latter bird is yet known by the few fragmentary remains which have been preserved , and that ornithologists need not hope ever to see better evidence of its former existence." ROBERT RIDGWAY

Edward`s Dodo. Early 1620s. Wikipedia Commons. Presented to the British Museum by George Edwards in 1759, having previously been in the collection of Sir Hans Sloane

June 25th 1874. p.311 June 25th 1874. **NEW SPECIES OF FISH IN IDAHO**

IDAHO – "An Idaho correspondent speaks of a new and undescribed species of fish that was last summer discovered to inhabit a small lake in the mountains, on a tributary of the Peyette River , weighing from five to ten pounds, in shape much like a shad, and having the skin and scales of a deep blood red color.

The meat is of a bright yellow color, and delicious to taste. Can you give me any information concerning the fish? Ans. Cannot be identified; description probably incorrect. May be a sucker."

Reference

1. Wikipedia http://en.wikipedia.org/wiki/Forest_and_Stream

The extracts here were taken from the set of *Forest and Stream* at the Biodiversity Heritage Library website - http://biodivlib.wikispaces.com/About

DISCUSSION DOCUMENT: DUTIES FOR REGIONAL REPRESENTATIVES

The CFZ had had regional representatives for over twenty years now. Some of them have done remarkable things, some nothing at all, and some something in between. I originally intended my first wife to manage the list of regional reps, but as history shows, that never happened.

Ever since Alison and I split up I have been intending to ask someone else to take over the job, and finally a few months ago I got around to it. Ronan Coghlan has agreed to take over the onerous task, and has come up with a list of suggested roles

for regional representatives, which I post here for public discussion.

[1] In the event of a reported sighting of a mystery animal in the representative's area, all possible data should be gathered and forwarded to CFZ. Likewise, news of further developments should be sent on as they occur.

[2] Representatives should try to discover if there were any sightings or other anomalous events in their areas in the past, but should only send on stories of UFOs or ghosts if they consider them important, as otherwise their is the danger of CFZ being swamped.

[3] Representatives should, if possible, look into local folklore to discover if stories of anomalous events in the area occur. Liaisons should be initiated with the Bird, Butterfly and Conservation Officer in their areas where possible. They should, in addition, try to gather an archive of Fortean zoological material from their local studies libraries.

[4] Representatives should initiate liaisons with groups dealing with anomalies and nature in the area, provided they consider them and their personnel suitable.

[5] Representatives should have the option of offering sales of books to local bookshops. However, some might find this distasteful and so this should not be regarded as an actual representative's duty.

The first weekend in April saw the very successful launch of Weird Weekend North: like the original Weird Weekend but up north, between Manchester and Liverpool.

The opening talk, by Richard Freeman, on dragons explored the nature of historical sightings and then explained that today nearly every dragon seen in film is not a true dragon but a wyvern (it is to do with the number of legs).

Steve Jones revealed that, far from being the stuff of film and legend, sightings of hooded entities appear more often than you think. One such sighting took place around a stone circle in Yorkshire that had been damaged due to fencing. Is it possible that the hooded entity is really a prehistoric guardian? Well if you'd been at Weird Weekend North you would have found out.

Perhaps one of the most surprising talks was the one regarding the recent increase in sightings of a UK Wildman, a mysterious bigfoot-like creature that is being reported more and more in our wooded areas. The witness accounts made interesting listening, especially as one well documented sight was

less than 15 miles away

When barrister Alan Murdie revealed in his talk that there is a stronger case in law for some poltergeist sightings based on the evidence than for some high profile criminal convictions it certainly make you think. A highlight of his talk was when a copy of a letter was produced and passed round; it was claimed to have been one of 5000 written by a poltergeist.

Just what are fairies? The talk looked at the nature of how people perceive fairies and also at belief in them. It was revealed that within living memory some people have been brought up to believe that they are fairies.

When Andy Lloyd presented the evidence for the Planet X it was a surprise to many just how much evidence there was for a large planet to be lurking out in the depths of space; it might not be just another planet but a failed sun, a twin to our own.

Lee Walker had the audience captivated with his stories from the heart of Liverpool. The tale of how a dog in a pub would predict the death of the older patrons had some in the audience in tears.

Award winning cartoon artist Hunt Emerson showed how closely linked his rise as a cartoonist was to that of the *Fortean Times* magazine. His work is collected worldwide and some people were lucky enough to get a very special caricature done by him.

Richard Freeman has recently returned from the forests of Tasmania where he has been looking for traces of the thylacine, the Tasmania wolf, an animal officially extinct since 1936.

Warburton souling play is a mummers play whose origins are lost in the depths of time, but during Laurence Armstrong's talk we learnt of the possible link to Eastern European plays. The soul caking play is unique to Cheshire and we are lucky enough to be close to Warburton and therefore able to see this unique play in action in early November.

After the previous day's poltergeist letter the tale of the Doddleston message brought us right up to the digital age. A basic computer in the 1980s started to receive messages from someone called Lucas, a man claiming to be from the past, and then folk from the future also started to get involved.

The whole weekend was a great success and Weird Weekend North will return in April 2017.

The editor and his compadres welcome letters for publication on all subjects covered by this magazine. However, we would like to stress that neither this magazine, or the CFZ are responsible for opinions expressed, which are purely those of the letter writer.

Dear Readers,

I was inspired by Jon Downes` very interesting essay in the CFZ 2016

GIANT EARTHWORM FOUND IN AFRICA

EAST LONDON, South Africa (AP) — Two visiting Swedish scientists teaching at the Transvaal's Potchefstroom University College, A. J. Reynecke and Perolof Ljungstrom, reported they found an earthworm 21 feet long on a road between Alice and King Williamstown, Cape Province. They said it was the largest ever found.

Yearbook on giant worms to look into the question of these creatures myself. I`m on the trail of these enigmatic animals, if you like. The incident of the largest of these animals in Africa, one that grew to be 21 feet long, was reported in The *Times Picayune* of November 12th 1967, a New Orleans newspaper.

I also found a rather poor quality photo in the Sunday Star (Washington D.C.) December 18th 1938 of the Gippsland ,Australia giant worm .If you think it looks like a piece of long rope then so do I ,but it gives an indication of its extraordinary length

This giant earthworm would furnish a lot of b angler. Earthworms that grow to a length of 10 fee in Australia's County of Gippsland.

An article titled Giant Earth Worms in the Philadelphia Inquirer of August 30th 1903 mentions this: " At the present time three especially large kinds of earth worms are known. One is found in South Africa,and another in *Ceylon and Southern India*, and the third is the one which has just been described (meaning the Gippsland worm – Richard) I have put some of the previous words in italics because these Indian and Ceylon (Sri Lanka) giant worms were not mentioned in Jon`s list in the Yearbook. Which is not meant to be a criticism, it just makes the giant worm situation more interesting. *The Morning Oregonian* of May 20th 1935 reported how the giant Oregonian worm, also called the night crawler, had been taken to Minnesota by fishermen to be used as bait.

Apparently a newspaper called the *Fertile Journal* of Minnesota reported that two Swedish fishermen, Hjalmar Erikson and Andrew Knutson had received some of these worms by air mail. One of these men tied the giant night crawler to a line and threw it out into the Minnesota trout stream resulting in the frenzy described here:

Trout large and small seemed to rush with the speed of lightning from all over the creek and straight for that worm. The water just boiled and churned, and after a minute or two when the line was lifted, lo and behold! there as the worm unharmed and hanging onto two of the biggest trout these men ever caught. The worm had wrapped each of its ends around trout's neck and choked the fish to death.

Another very peculiar thing I came across is the story of the 20ft long bulge that appeared for no reason in a Fort Worth, Texas street.

It was "like a giant earthworm" trying to surface then disappeared without a trace. None of us have ever seen anything like it before" said R.J. Forester, a Fort Worth Fire Department official. "It grew for about an hour and then we just turned our heads and it was gone". As recorded in the U.S. Newspaper *Centre Daily Times* July 16th 1984.

The *Times-Picayune* (New Orleans, Louisiana) June 18TH 1916 reported a worm that could squirt, discovered in the Belgian Congo. The Belgian Congo is now the Democratic Republic of Congo.

A Huge Earthworm.

A curious new earthworm found by Dr. Cuthbert Christy in the Belgian Congo may be a little more than a foot long, but its striking peculiarity is that of squirting. From this H. A. Baylis has named the species dichogaster jaculatrix. The jets of slightly viscid fluid are thrown out when the animal is irritated, and the first discharge may be followed by a second within a few minutes, and sometimes even by a third. They may reach a height of ten or twelve inches. The cuticle of the living worm shows a bright green and blue iridescence, and a purple band marks the back of each segment. The red chimney-like casts sticking up four or five inches from the worms' burrows in red clay are sometimes baked for tobacco pipes by the native miners.

http://cri.csyangsheng.com/detail/qwbg/yixingguaiwu/99420/7

This Chinese language web site shows what appears to be a giant worm in a ditch.

Best Wishes

Richard Muirhead

Reviews

SUZANNE SELFORS

THE SASQUATCH ESCAPE

THE IMAGINARY VETERINARY: BOOK 1

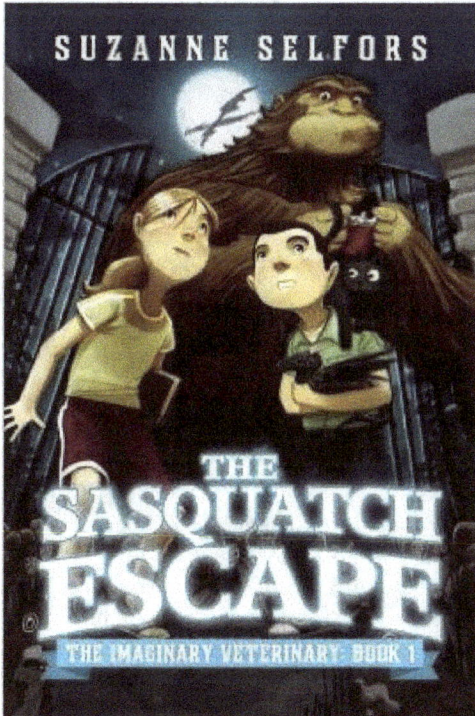

Paperback: 214 pages
Publisher: Little, Brown Books for Young Readers (7 Jan. 2014)
Language: English
ISBN-10: 031622569X
ISBN-13: 978-0316225694

Regular readers of my writings across the board (and I am always both impressed and surprised to find the number of people who read my writings on both cryptozoology and rock and roll, and other subjects in between) will remember my friend Elizabeth Clem. I have mentioned her on occasion over the years, because she sends me a wide selection of interesting stories, and always has, even when she was not much more than a schoolgirl.

Our paths first met when I was researching lake monster stories from Indiana, and she did most of the work for me. Then in 2004 I was in Illinois chasing up stories of black panthers, and she and her family crossed state boundaries to come and meet me. And we have been friends ever since.

When Corinna and I got married in 2007, Elizabeth sent us a whole parcel of books as a wedding present, and these included *Cryptid Hunter* by Roland Smith, who also then became a mate, quite possibly because he admitted that he had based one of the main characters in the book on an idealised version of yours truly (but without the epic bad behaviour which had marked so much of my earlier life). And now, nine years after that I am reviewing another piece of cryptoliterature that arrived on the recommendation of the delightful Ms Clem.

Whereas *Cryptid Hunter* and its four sequels were aimed at a young adult audience, with a modicum of violence and romance, these books (and although I have only read the first of them, so far, there are now six or seven in the series) are aimed at a younger tween audience. But this is not to say that they have less to recommend them.

In fact this is a surprisingly sophisticated little book, which works on a number of different levels. Just deconstructing the back cover blurb delivers some surprises. Let's have a go:

"When Ben Silverstein is sent to the rundown town of Buttonville to spend the summer with his grandfather, he's certain it will be the most boring holiday ever. That is, until his grandfather's cat brings home what looks like . . . a baby dragon?

Enter Pearl Petal, a local girl with an eye for adventure, who helps Ben take the wounded dragon to the only veterinarian in town - the mysterious Dr Woo. No one knows where Dr Woo came from or why she's moved into the old button factory and renamed it Dr Woo's Worm Hospital. But as Ben and Pearl discover once they are inside, Dr Woo's isn't a worm hospital at all - it's actually a secret hospital for Imaginary Creatures.

After Ben accidentally leaves the hospital's front door unlocked, a rather large, rather stinky, and very hairy beast escapes into Buttonville. Ben and Pearl are tasked with retrieving the runaway creature, and what started out as the most boring holiday ever becomes the story of a lifetime. . ."

OK I come from a very different generation, and the children's books of my childhood very seldom touched on social and family problems. But my wife and I were both married before, and I know full well that kids' books these days do address the fact that 'happy ever after' quite often doesn't work out the way it was planned. But I have seldom read family problems dealt with so sensitively as they are in this book. Ben is staying in Buttonville because his parents are having troubles and they need time without him to sort them out, but this is dealt with gently, sensitively and without labouring the point like so many other books of this type do. But it is what happens when he arrives in Buttonville that is so interesting. Because Buttonville is the sort of conurbation that seems to be coming ever more prevalent; a place where the main industry and source of employment has closed down some years before. As a result of this the population of working age has largely moved away, and one is left with a population that is largely old, and that is growing older and more infirm with every year.

It is dealt with in an amusing way suitable for children, but there is no denying that a town where there is nothing to do but go and eat pudding at the Senior Citizen's Centre is a desolate place for a ten year old. As someone who, in a few short years, will be eligible for the Over Sixties Club, it sounds rather nice, but I am not part of the chosen demographic that this book is aimed at.

Ever since its inception a quarter of a century ago, the Centre for Fortean Zoology, the organisation I founded, has dealt with creatures dwelling in the grey area between Natural and Unnatural History; the place where science and mythology meet. And it is admirable, at least as far as I am concerned, that this book operates in much the same area. The Fortean phenomena which the protagonists experience earlier on in the story take place in the corner of one's eye, and behind drifting clouds, and - especially considering that it is a humorous book aimed at pre-teenage children - it has a very convincing sense of mystery about it.

I have always said that many of the better known animal mysteries of the 21st century - such as bigfoot, the thylacine, and British mystery cats - will only be solved conclusively when one of them is hit by a car, and becomes a victim of the sort of random act of violence for which it is impossible to legislate, and so I was rather pleased when the two young protagonists' first real brush with the unknown is when a baby dragon is brought into the house by the old man's moggy. This is the way that the universe seems to work, and - despite the tone of the book being aimed firmly at the younger reader - much of this mirrors how quasi-Fortean investigations actually take place, at least in my experience.

But I think that what I like most about this book is the fact that at the end there are a series of bonus writing, art, and science activities that will help readers discover more about the mythological creatures featured in *The Sasquatch Escape*. These activities are designed for the home and the classroom. And the readers are urged to enjoy doing them on their own or with friends! I cannot recommend this little book highly enough and look forward to reading the sequels in due course.

Well done Ms Selfors. JD

The Jungle Book
Disney 2016
Directed by Jon Favreau

Ask the average person in the street about 'The Jungle Book' and they will almost certainly think of Disney's 1967 cartoon with its singing animals. Few people have read Rudyard Kipling's two collections of short stories *The Jungle Book* (1894) and *The Second Jungle Book* (1895). Kipling wrote the books for his daughter Josephine, who died in 1899 aged six. The books are a dark and wonderful fantasy invoking darkest India where nature is red in tooth and claw, and the wild places are still untamed. A theme that runs through the books is an honour among beasts and a resistance to the destructive ways of man. 'The Jungle Books' are truly some of the finest pieces of literature that the UK has ever produced.

It is all the more enraging that Disney chose to cheapen and bastardize Kipling's work in his wretched cartoon. Walt Disney famously advised his animators not to read the original books so that they would not be influenced by the dark tales therein.

Instead we have an infantile pantomime in which Kaa the python, Mowgli's friend and oldest and wisest teacher, who rescues Mowgli from the Cold Lairs when he is kidnapped by the monkey tribe, or Bandar Log, is transformed into a cowardly villain playing second fiddle to Shere Khan. The reason? Walt Disney thought that right wing Christians in the USA would never accept a snake in a heroic roll.

To add insult to injury Disney added new characters to the story including King Louie the orang-utan (not native to India) as the king of the Bandar Log (who by rights have no king).

Even the animation is poor and all the darkness and wonder of the originals is lost in a tide of saccharine nonsense. Walt Disney himself also deliberately

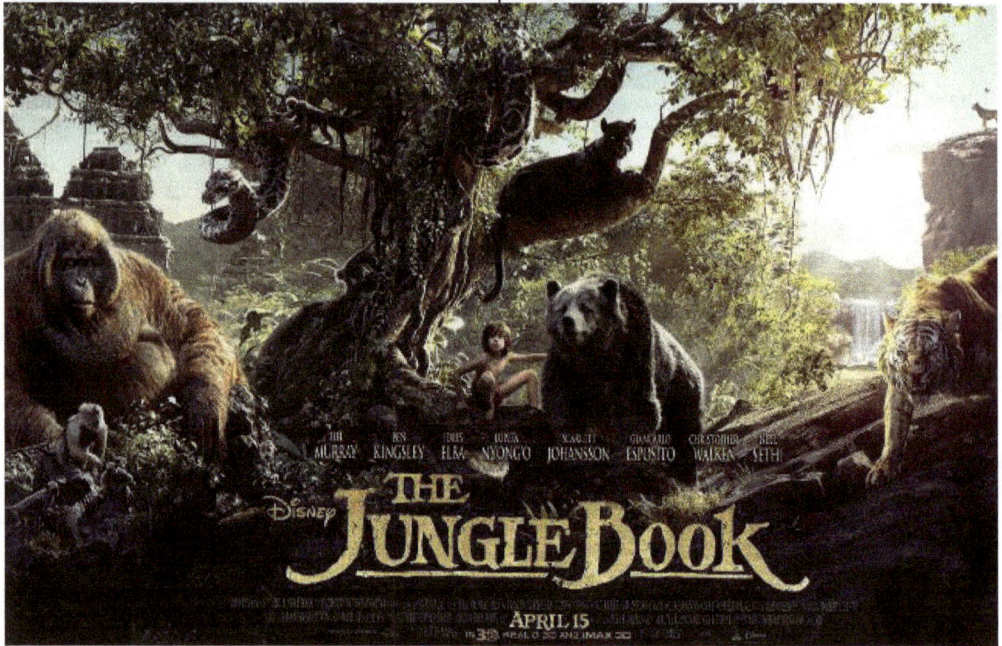

mispronounced Mowgli's name.

All of the above are unpardonable crimes against literature. Disney seems to have a flair for raping classic British books. They also produced an execrable animated version of T.H. White's *The Sword in the Stone*.

Not satisfied with this, the odious Disney corporation have returned again and again to besmirch Kipling's books with a live action 1994 version that also makes Kaa a villain and includes the spurious King Louie. In this version the animals are struck dumb.

In 1998 an even worse film was released by Disney studios direct to video. *The Jungle Book: Mowgli's Story* features African chimpanzees as the Bandar Log (who live in 'monkey town rather than the Cold Lairs) as well as having African

baboons, South American macaws and a North American skunk in it.

Unbelievably, in 1990, Disney developed a cartoon series called *Talespin* that cast Baloo the bear, another of Mowgli's venerable teachers, as a 1930s bush pilot in an Indonesia type setting with Shere Khan as a corrupt business mogul. Yes, you read that correctly, and no I've not been drinking.

The height of stupidity was reached in the 1996 *Jungle Cubs* that re-imagined the characters from the 1967 animated film as youngsters and featured a hip-hop version of the '67 film's song 'The Bear Necessities' ... Jesus Christ on a god damn bike.

Well I've procrastinated over Disney's past crimes long enough. How about their latest offering?

and Bill Murray as Baloo.

On the downside Baloo is shown as a rather lazy and dim-witted bear rather than the wise tutor of the books. Again Kaa (voiced here by Scarlett Johansson) is once again a villain who attempts to devour Mowgli.

The anachronistic King Louie turns up, this time rationalized as a *Gigantopithicus blacki,* a monster ape known from fossilized teeth and jaw bones found in India, China and Vietnam, the latest of which are some 300,000 years old. Voiced by Christopher Walken he is here depicted as an outsized Bornean orang-utan (*Pongo pygmaeus*) rather than the erect walking putative yeti we are more familiar with.

Well it is certainly a triumph of special effects. All of the thousands of animals on screen are CGI and very well rendered too. The animators must be congratulated for their attention to detail. Early on in the film there is a scene where all the jungle animals are crowded around a drinking hole during a drought. In the crowd can be seen some pink headed ducks (*Rhodonessa caryophyllacea*). This is a species with cryptozoological connotations. Believed extinct since the 1950s, reports suggest that it may still be hanging on in the wetlands of northern Burma. British cryptozoologist Richard Thorns has taken a number of expeditions into the area in search of them.

Mowgli himself is played by Neel Sethi who puts in an impressive performance. Considering he was acting against nothing, the CGI animals created later, he really does make a believable Man Cub. The voice cast is impressive with the charismatic Idris Elba suitably menacing as Shere Khan, Ben Kingsley as Bagheera

Whilst being far, far superior to other Disney efforts over the years, this version of *The Jungle Book* falls far short of capturing the dark majesty of Kipling's books. I would direct interested viewers to Zoltan Korda's 1942 version film *Jungle Book* staring Sabu Dastagir as Mowgli. This version treats the source material with the upmost respect and evokes the books in a way no other film has done since.

As I write, the fine actor Andy Serkis is working on a Universal Pictures version of *The Jungle Book*. Directing and playing Baloo he will be joined by Benedict Cumberbatch as Shere Khan, Christian Bale as Bagheera and Cate Blanchett as Kaa.

Without the yoke of Disney and the long, childish shadow of the '67 cartoon, we may see the finest version of Kipling's timeless classics since the 1940s.
3/10 RF

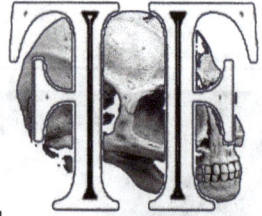

THE WORLD'S WEIRDEST PUBLISHING GROUP

We publish a lot of books. Indeed, I think that we could quite easily claim to be the world's foremost publishers of books about Fortean Zoology and allied disciplines, and our Fortean Words imprint is doing a great job in producing books on other non-zoological esoterica. However, I feel that it would be unethical to review our own titles. So here, to end this edition of *Animals & Men*, is a brief look at the books we have put out so far this year.

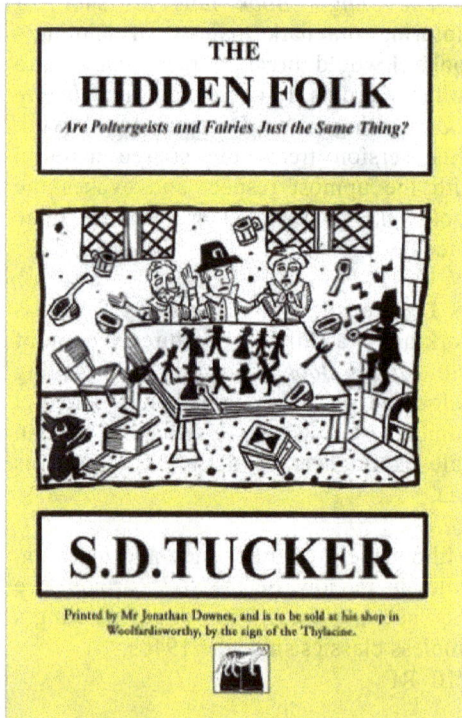

THE HIDDEN FOLK

Are Poltergeists and Fairies Just the Same Thing?

S.D. TUCKER

Printed by Mr Jonathan Downes, and is to be sold at his shop in Woolfardisworthy, by the sign of the Thylacine.

Paperback: 354 pages
Publisher: Fortean Words (13 Mar. 2016)
Language: English
ISBN-10: 1909488402
ISBN-13: 978-1909488403

Fairies were genuinely believed in right the way across Europe in the not-too-distant past, and not only by little girls. For many adults, fairies were a part of everyday reality, and accounts of their alleged interaction with the world of humans are legion. But, when the 'realm of faery' did intrude upon our own, how did its inhabitants make their presence known? Apparently, they did so in a variety of different ways; by rapping, tapping and making loud noises around a person's home, by throwing pebbles about, by setting fires, moving furniture, breaking plates and speaking from out of thin air itself. In other words, then, they acted just like poltergeists are often said to do today.

2016 YEARBOOK

Edited and Compiled by
Jonathan and Corinna Downes

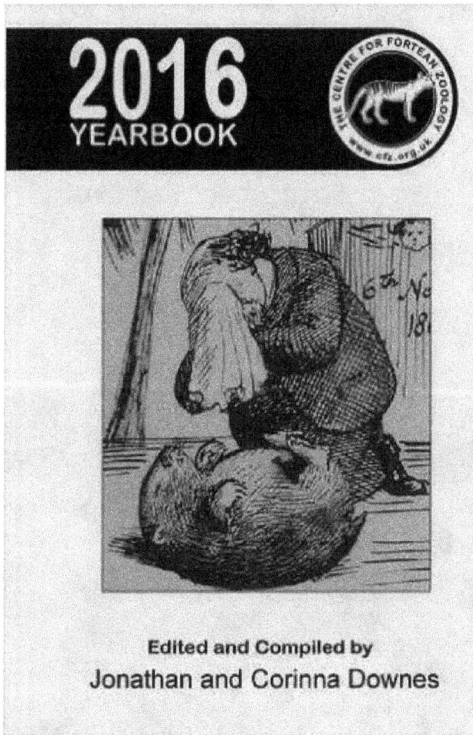

Paperback: 210 pages
Publisher: cfz (11 May 2016)
Language: English
ISBN-10: 1909488429
ISBN-13: 978-1909488427

The Centre for Fortean Zoology Yearbook is an annual collection of papers and essays too long and detailed for publication in the CFZ Journal, Animals & Men. With contributions from both well-known researchers, and relative newcomers to the field, the Yearbook provides a forum where new theories can be expounded, and work on little-known cryptids discussed.

CONTENTS

www.ingramcontent.com/pod-product-compliance
Lightning Source LLC
Chambersburg PA
CBHW050552280326
41933CB00011B/1813